Guinea

Past and Present

by

Michael Roberts

Edited and Illustrated by

Sara Roadnight

Photography by

Michael Roberts

Cover photograph
*Lavender guineafowl
and Crested guineafowl*

2002 Michael Roberts.
Published by Gold Cockerel Books
ISBN 0 947870 36 9

Printed by Bartlett & Son Printers
Swan Yard, St Thomas. Exeter.
Tel. 01 392 254086

Acknowledgements

Many thanks must be given to the following people and organisations for their assistance, without which this book would not have been possible.

Birdworld, Farnham, Surrey

Bodleian Library, Oxford

British Museum, London.

British Museum, Tring.

Mrs Boundy of ukguineafowl

Cairo Museum

David Rockingham-Gill, Zimbabwe.

Exmoor Bird Gardens.

Galor, France

Jean Champagne, CIP. France.

Jean Olliver, SAVEL. France

Julian Bird, Egyptologist

Linnean Society.

Mattia Buondonno, Pompeii

Museo Nazionale Romano, Rome

Nigel Strudwick, British Museum

Ralph Winte, USA

Reading University.

Terme di Diocleziano, Rome

Winchester Cathedral Library

Introduction

This book has been written partly in response to various people asking about the viability of rearing guineafowl for the table. Some of these people are large pheasant rearers who feel that, due to political pressures, their customers could be restricted in the number of pheasants they are allowed to release on their shoots.

One of the problems is that while the British eat pheasant, they rarely eat guineafowl, which is odd when you consider our former Colonial connections with the West Indies and East and South Africa where guineafowl is often on the menu.

In France today there are over 700 small producers of guineafowl for the table; small in France means between 2000 and 3000 birds per annum, while there are many rearing between10,000 and 15,000 birds per annum, and still only 3% of the French eat guineafowl! So it is a question of education and marketing, both of which we will be looking at in detail later on.

This book covers all aspects of keeping guineafowl, whether for pleasure or commercially. We hope you enjoy reading it.

Michael Roberts and Sara Roadnight
Kennerleigh, Devon.

May 2002

Contents

THE HISTORY OF GUINEAFOWL

The history of guineafowl is incomplete and inexact as yet. Chicken and guineafowl bones are very similar to the untutored eye, but these days new specialised knowledge of bird bones is helping to shed light, not only on existing specimens, but also on new finds; as a result the history of guineafowl is now developing and evolving year by year.

The first encounter with these birds in Europe (Mlikovsky 1986) took place in France when a leg bone dating from the late Eocene period (35 - 55 million years ago) was found near Lalbenque in Lot. A guineafowl fossil from the same period was also found in China (Wetmore 1934, Olson 1974).

The first known mention of guineafowl is in Egypt and is to be found on the East Gable in the pyramid of Unas at Saqqara. The date is about 2,400 B.C. The passage which includes the guineafowl hieroglyph comes in the so called Cannibal Hymn. Unfortunately this pyramid was closed to the public when we were there.

The earliest mention of guineafowl in literature comes from the Greek myth of the killing of the Calydonian Boar. Oeneus the king of Calydon forgot to include Artemis in his yearly sacrifice to the gods. When she found out she sent a large boar to ravage Calydon. Oeneus invited the bravest hunters to try to kill it, with the promise of its skin and tusks as a reward. There was a great fight with this boar involving the huntress Atalanta and a number of other hunters. During the skirmish several of the hunters were killed, but Atalanta fired the first arrow to hit the boar behind the ear, and another hunter, Meleager, speared it in the flank which eventually killed it. Meleager, who was in love with Atalanta, said that the skin and tusks should go to her as she had drawn the first blood, but the other hunters disputed this, not really approving of a female in their midst. Meleager, in a lover's rage, killed these two hunters, but was then killed himself. Artemis turned all but two of his sorrowing sisters into guineafowl, the white spots on their plumage representing their tears. From Meleager we get the Latin names Meleagris and Meleagrides.

The ancient Egyptians were great keepers of birds of all kinds. The guineafowl was not only depicted in their tombs but was also part of their written language, and appears in varying forms as the hieroglyph NH, pronounced 'narh'. We were hoping to come across this hieroglyph without too much trouble while we were at Saqqara and the Valley of the Kings. You would have thought that the words 'pray for' or 'lotus bud' or 'eternity' for example, which all contain 'nh', would certainly have occured from time to time, but we didn't spot them at all. Perhaps a simpler variation was used instead.

We were more successful at Karnak however. This vast complex lies one and a half miles north of Luxor and it is here that the ruined White Temple of

𓃹𓏭𓅮 *nḥ* (G 21) guinea-fowl.

𓃹𓏭𓀢 var. 𓏴𓏭𓀢 *nḥt* (G 21) pray for (something) ; *nḥ*, *nḥt* prayer.

𓏏𓏲𓃹𓏤 *nḥs* (be) hard, rough, dangerous.

𓃹𓃀𓏴 *nḥb* yoke together, unite; equip, *m* with ; *Nḥb-kꜣw* det. 𓆓 (D 30) Uniter-of-attributes, name of a mythical serpent; det. 𓆙 Neḥeb-kaw, feast of the month later called Khoiak, see p. 205.

𓃹𓃀𓄹 *nḥbt* neck.

𓃹𓃀𓄻 *nḥbt* (M 10) lotus bud.

𓃹𓃀𓏐 *nḥp* potter's wheel.

𓂝𓃹𓅓 *nḥm* take away, rescue, *m-ꜥ* from (someone); *Nḥmt-ꜥwꜣy* She-who-rescues-the-robbed, consort of the god Thoth at Hermopolis.

𓂝𓃹𓏝 *nḥmn* non-encl. part., surely, assuredly, §§ 119, 6; 236.

𓃹𓏭𓇳 var. 𓍿𓇳 *nḥḥ* eternity.

𓃹𓏭𓈈𓃠 var. 𓏭𓈈 *Nḥsy* (T 14) Nubian.

𓃹𓏤𓂧 *nḥdt* tooth, molar; see too *ndḥt* below.

Sesostris 1 (1971 - 1926 B.C.) has been reconstructed in the Open Air Museum to the north of the Great Court. We were searching for a specific relief of a guineafowl about 4inches high, one of the many hundreds of hieroglyphs that covered every surface in the temple from floor to ceiling. We very nearly missed it but found it in the end on the third column on the right facing the front at eye level! The hieroglyphs read downwards from the right and say: "I (meaning the king) made a beautiful monument, (I) gave (it) to you (Amun-Re) for eternity." In this text the guineafowl hieroglyph is part of the word 'nhh' meaning eternity.

Across the river Nile from Luxor, on the West Bank lie the Valley of the Kings, the Valley of the Queens and the less well known Tombs of the Nobles. There is a painted hieroglyph of a guineafowl in the tomb of Senneferi or Sennefer (TT99) who was an important builder around 1420 B.C. This hieroglyph appears on a pillar at the back of the tomb where Sennefer is shown receiving New Year gifts from his wife Taiamu, his children and some craftsmen.

The Ancient Greeks knew about guineafowl from their various travels round Africa; in New York at the Metropolitan Museum of Arts there is a skyphos or type of vase on display dated 600 B.C. with guineafowl depicted on it in black. Aristotle (448 - 380 B.C.) writes: "The colours of eggs vary in different kinds of birds. Some have white eggs as pigeon, partridges, some yellow as those inhabiting streams; others are spotted as those of the meleagris and phasianus."

Marcus Terentius Varro (116 - 27 B.C.) writes: "the African or Numidian hen is different as those the Romans call gibberae (humped back)" and "the African hens are large, speckled with rounded back, and the Greeks call them Meleagrides. These are the latest fowls to come from the kitchen to the dining room, because of the pampered tastes of the people. On account of their scarcity they fetch a high price."

Quintus Horace (65 - 8 B.C.) wrote in the Epodes (29 B.C.) : "Not Afric fowl, not Ionian pheasant would make for me a repast more savoury than the olives gathered from the richest branches of the tree."

Pliny the Older (23 - 79 A.D.), while writing about guineafowl says: "The Meleagrides in Boeotia (an area just north of Athens) fight in similar manner; this is a kind of hen belonging to Africa, hump backed and with speckled plumage", and "in the Numidian port of Africa (roughly Algeria) the Numidic fowl, all these are now found in Italy."

Columella, writing in the same century, speaks about "the African fowl, (which he wrote as Africana) which most people call Numidian, resembles the Meleagris, except that it has on its head a red helmet and crest, both of which are blue on the Meleagris."

By this time the Romans had come up through France to England bringing with them all sorts of birds including guineafowl. We know this from finds at various sites, and there is still a lot of work being carried out into the early imports of these birds. A leg bone of a guineafowl was thought to be among the 17 species of bird bone found in excavations at the old Roman town of Calleva Atrebatum (Silchester). This has not been verified for sure. It does seem strange however, that although there were chicken bones, which are very similar, found at the same site, someone was convinced they had come across something different. Furthermore, the leg had a metal ring or tag attached to it, which raises another interesting historical point: perhaps this was the first recorded discovery of a bird that had been ringed in Britain.

Other curious finds at Silchester were raven and crow bones which suggest that these birds may have been kept as pets. Perhaps they were used to warn off intruders or to welcome invited guests. Magpies were kept in barbers' shops in Rome. (Is there some connection between chattering magpies and barbers?!) Could the guineafowl, (?), with the ring on its leg also have been used as an alarm bird?

It has always been understood that guineafowl died out after the Romans left Britain. They only reappeared in the 14th and 15th centuries when explorers began to bring back new discoveries from far flung places around the world. But there is an interesting reference in Kennett White's Parochial Antiquities of Oxfordshire and Berkshire, 1695, which seems to refute this. Several copies are to be found in the Bodleian Library in Oxford. The relevant passage comes from part of a record written in 1277 concerning Ambrosden and Burchester, now Bicester, in Oxfordshire. It is written in Latin and concerns, "one John (Johannes) Willard who holds in a granary 5 quarters and a half of corn (frumenti), 4 oxen (boves), 6 sheep (mutilones) and 6 female guineafowl (Africanae faeminae)." There are conflicting notes on the translation of Africanae, (Columella uses this word) but one note says: "A certain number of this sort of fowl was frequently reserved among the provisions paid to the Lord (of the Manor) from his customary tenents." Bearing in mind that this is a record that only concerns Oxfordshire and Berkshire, one can't help wondering how many other guineafowl there were in England at that time. It's possible that there were as many as there are today, scattered around the country in small semi-domesticated flocks. Of course the fox was comparatively rare in those days.

When Marco Polo, (1254 - 1324) was travelling in the Province of Abyssinia he wrote that , "They have the prettiest hens to be seen anywhere." It's possible that he was referring to Vulturine guineafowl.

With the opening up of the New World and the era of exploration, guineafowl began to increase in numbers. They made a nice addition to the bird collections of the 'new rich', along with peacocks, swans, shelduck and other exotic breeds.

Pierre Belon, (1517 - 1564) who used the name 'poule de la Guinee' for guineafowl, wrote in his Histoire des Oiseaux: "As many of the products which merchants bring to France from Guinea were at first unknown, hence these hens were familiar to none of us before navigation to that region began, but now they are prominent enough and common in the residence of rich men....... There is no clear method of distinguishing the male from the female....... their cry is similar to that of common hens, for they call out sharply in a high voice like that of recently hatched chicks. They sit on hen perches as our hens do ; their flesh is delicate and their eggs suitable for eating." (I wonder whether Belon really studied the live birds; their sexes are in fact very easily distinguished by their calls. He certainly enjoyed eating them!)

In Willughby's Ornithology, (1678) translated by Ray, there is a reference to a text from the German Gesner: " the Mauritanian cock is a very beautiful bird, in bigness and shape of body, Bill and Foot like a pheasant........ armed with a horny Crown, rising up into a point, on the backside, perpendicularly, on the foreside with a gentle ascent or declivity......... so that it fits on the head after the same manner as the Ducal Cap doth upon the head of the Duke of Venice."

A century later Georges Louis Buffon, (1707 - 1788) records the guineafowl as 'la peintade' in his Histoire Naturelle des Oiseaux, and writes: "The guineafowl which were raised so carefully in Rome, were lost to Europe (after the Roman era) as there was no mention of them by the writers in the Middle Ages and that they only became talked about since the Europeans frequented the West coast of Africa......." The priest Charlevoix affirmed that: "There is a type of guineafowl at Santo Domingo (on the island of the Dominican Republic), smaller than the ordinary type but appears to be red. It arrived there with the Castillians." (This is interesting as there is talk in the United States at the moment of a red guineafowl but nobody has seen one yet!). "At the beginning of this century (1700), guineafowl were again very rare in England".

In Lewis Wright's Book of Poultry (circa 1860), these birds are mentioned: "The early guineafowl would have reached Egypt from Abyssinia, and could have been the forerunner of our European stock............ At Bristol, which is a considerable centre of the West African trade, we have on many occasions seen guineafowl perched on the rigging of African vessels, brought from the coast by sailors; and in every case those were obviously identical with the domestic breed, both in head and plumage, being only somewhat slighter in build......... Hybrids have occured between the guineafowl and common poultry. We knew of one with a Dark Brahma cock." (This is interesting as I once bred a guineafowl / Light Brahma cross by accident. See the photograph.)

In Wingfield and Johnson's The Poultry Book (circa 1860), there appears: "Hybrids between the guineafowl and common fowl are said to occur. About three years since a couple of these monsters were alleged to be in the aviary of the London Zoological Society."

In his book Bird Behaviour (circa 1930), Frank Finn writes: "The guineafowl has varied but little, only showing some colour abberations as a rule, though I once got one in India that would have delighted Darwin, as it had a pendulous tuft of feathers hanging from its neck, much like the Turkey's beard of bristles, which Darwin said would have been called a monstrosity had it appeared under domestication." (This is rather interesting as some of the Egyptian hieroglyphs portray a bird with a tassle on its neck.)

NAMES OF GUINEAFOWL

The name 'guinea' comes from the West African coast of Guinea which used to extend from present day Senegal to as far south as Gabon. It derives from the Tuareg word 'aginaw' which means 'black people'. It was this 'guinea' that gave the name to the gold coin from that area.

There are various regional names for guineafowl such as guinney, gynney, ginny or gleeny. In the West of England people use the names galliny or galeeny which may originally have derived from gallina, a hen in Latin. Shakespeare speaks of a Gynney hen in Othello

The word keet means a young guineafowl and is sometimes used for day-olds as well. It comes from the old Nordic word cytling or keetling, the ending 'ling' being a diminutive meaning small as in gosling. (I'm not sure if the old Nordic people ever came across guineafowl though.)

In ancient times, the Romans used to write about the galina numidia or the numidian hen. Numidia was roughly where Algeria is today. The name then changed to Aves africae or Africanas meaning African birds or Africans. The Ancient Greeks used the word Meleagris taken from the myth about Meleager and his sisters. In classical Arabic the name used is ghergher, but in Egypt today the bird is called a dagag or farrouge el wadi, chicken of the valley.

When it comes to other countries the name for guineafowl varies considerably. In Portugal it is called the gallena pintada which is interesting because in France the bird used to be called la poule de Guinee (the hen from Guinea), which later changed to peintade. The word peindre means to paint in French, and the modern name for guineafowl in France is pintade. The Italians and Greeks use the name faraone, pharaoh bird, and the Germans, perlhuhn, pearl hen.

THE ORIGINS AND TYPES OF GUINEAFOWL

Guineafowl are part of the pheasant family as we can see from the diagram on the opposite page.

Guineafowl are divided into four groups. The domestic bird comes from the second group or genus called Numida or Helmeted guineafowl.

1) Group or genus Agelastes of which there are two species:

Agelastes meleagrides or Whitebreasted guineafowl
Agelastes niger or Black guineafowl

	Distribution
Agelastes meleagrides or Whitebreasted guineafowl	Sierra Leone, Ivory Coast, Ghana
Agelastes niger or Black guineafowl	Nigeria, Angola, Congo

2) Group or genus Numida of which there are nine species:

Numida meleagris coronata	South Africa,
damarensis	Angola, Botswana, Namibia
galeata	Central & West Africa,
maringensis	Central South Africa,
meleagris	Central East Africa,
mitrata	Mozambique, Zambia, Tanzania,
reichenowi	Tanzania, Kenya
sabyi	Morocco
somoliensis	Somalia, Ethiopa

3) Group or genus Guttera, two species:
Numida plumifera, two sub-species:

Numida plumifera plumifera	Eastern Central Africa
Numida plumifera schubotzi	Zaire

Numida pucherani, five sub-species:

Numida pucherani edouardi	Mozambique, Zambia
barbata	Malawi, Tanzania, Mozambique,
pucherani	East Africa,
sclateri	Cameroon
verreauxi	Ivory Coast

4) Group or genus Acryllium of which there is only one species:

Acrillium vulturinum	Ethiopia, Somalia, Kenya

Guineafowl have been introduced into many parts of the world but without much success, due to predators and adverse climate conditions. Although there are semi-domestic flocks to be found in many parts of the world, they are not truly feral except in parts of the Carribean, Madagascar and South Africa, (see "A Letter from Zimbabwe").

It is an interesting fact that it is only in the last 70 years that guineafowl have been commercialised, mainly by the French and the Italians.

AVES

GALLIFORMES

Megapodes Guans Grouse Pheasants & Quails Guineafowl Turkeys

mitrata

coronata

galeata

meleagris

reichenowi

pucherani

vulturine

schubotzi

A LETTER FROM ZIMBABWE
BY DAVID ROCKINGHAM-GILL

The guineafowl that occurs in Zimbabwe is Numida meleagris coronata. The males weigh approximately 1475 grams (3 1/4 lbs) and the females about 1492 grams (3 ⅓ lbs), the males being lighter than the females.

In the Atlas of South African Birds guineafowl were not found south of the Orange River before the twentieth century, but a combination of factors such as cultivation, expansion of water points for cattle and sheep, telephone poles to roost on etc., have taken them all the way to Cape Town.

These birds occur from our lowest forest at 300 m to about 1800 m, but they do get sparse in the mountainous country in the east.

Critical habitat features are roost sites close to water and cover. Short grassland is best but they will go into long grass when pressed. They like to put their heads above the grass and have a look all the time. They like to roost in trees and a Msasa tree is good; their perches are the bigger branches below the canopy, obviously a comfortable fit for their feet.

Guineafowl have been pushed out of the major towns by dogs, cats and people, but I saw some on the Police golf course the other day, some 4 km from the centre of Harare.

In the area I live, Mashonaland West near Chinhoyi, I have seen flocks of guineafowl up to 300, but the norm is more like 20 to100 birds.

There is an article in the most recent "Birds and Birding" about local extinctions, and this deals with some place in Natal. Quite why guineafowl have gone from there I don't know, but in Zimbabwe we do get local extinctions from time to time and they seem always to be after an outbreak of Newcastle Disease.

There is little credit to guineafowl in many books I have read on South African history, but from the Great Trek and in the days of the ox wagon, the wagons followed the roads made by others and the game along these roads was very soon shot out, therefore the guineafowl and francolin must have been shot for the pot all the way along the route to keep people in relish. Even I have shot them for the pot, many years ago when opening up a farm, living in a tent and going out with a shotgun to bag something to eat.

Then the motor car came along and they must have been shot even more until people got fridges and were able to store food, when the onslaught must have let up, and now with more waste on the farms, the guineafowl are doing extremely well.

Of course this depends on the season, in years of drought they don't fare as well, mostly because of visibility as they get hammered by the predators, avian raptors who can see them as there is no hiding place, jackals, servals, baboons and monkeys. Mongeese, rats and snakes find their eggs very easily. So almost anything will eat guineafowl and their mortality must be very high, probably 95%.

Guineafowl do get knocked over on the road. I have been in a car where the windscreen was broken by the impact, and they can do considerable damage to headlights and panels.

Their flight is powerful, swift and direct and they are easily shot with a gun. They can fly up to a kilometre, but in a normal day they only fly up to their roost site and down to the ground.

We have always shot Helmeted Guineafowl but when I was a boy the Crested Guineafowl was Royal Game, stupidly so because they are not uncommon where they live, sometimes together with Helmeted Guineafowl but in different niches in the Zambezi and Sabi-Limpopo valleys. The Crested Guineafowl is much more independent of water.

For the pot I try and shoot the youngsters with grey heads because the blue headed jobs are pretty tough. They do not have an ounce of fat on them and therefore the best thing to do is to inject their meat with melted butter before and during cooking.

They pair off in the hot wet season, (November to April) then take to the bush where the males do their fighting and jostling for females. The first eggs are laid in November / December, broad at the top and pointed at the bottom and pitted all over. Clutch sizes are between 6 to 15, more if two females are using the same nest. The shells of these eggs are possibly resistant to grass fires. Nests are often under a fallen thorn tree on the ground and are quite difficult to find, initially a scrape which gets deeper with incubation. Eggs are laid on a daily basis and only the female sits on them. Depending on the weather pattern, there can be several clutches through until April, but the birds start to flock again as soon as the chicks hatch. Snakes must take quite a few eggs.

The female incubates the eggs and the male does 80% of the brooding when the chicks have hatched. They are precocial, (born with eyes open and bodies covered with down) and can feed themselves without trouble after 24 hours. They can fly short distances after two weeks. Some nests have a few infertile eggs or at least the eggs never hatch, (probably due to two hens sharing the same nest. M.D.LL.R.) and the nests are often found in the cold dry season, long after the chicks have gone.

They readily feed in the open and in fact often gather in the open, possibly for other birds to see them so that they can accumulate into a flock.

They eat mostly protein. The crop of one bird collected in the late evening contained 243 grams of harvester termites, Hodotermes mossambicus, estimated at 5,100 individuals. They therefore help all mealie (Zea mays) farmers by reducing the white ant population and reducing lodging. I have shot hundreds on a Syndicate shoot and yet we do no damage to the population at all. I have inspected the crops of those I have been given and they have had very few mealie pips (seeds) in them. They do well on commercial farmland and particularly like soya beans which are dropped or not reaped by the combine harvester.

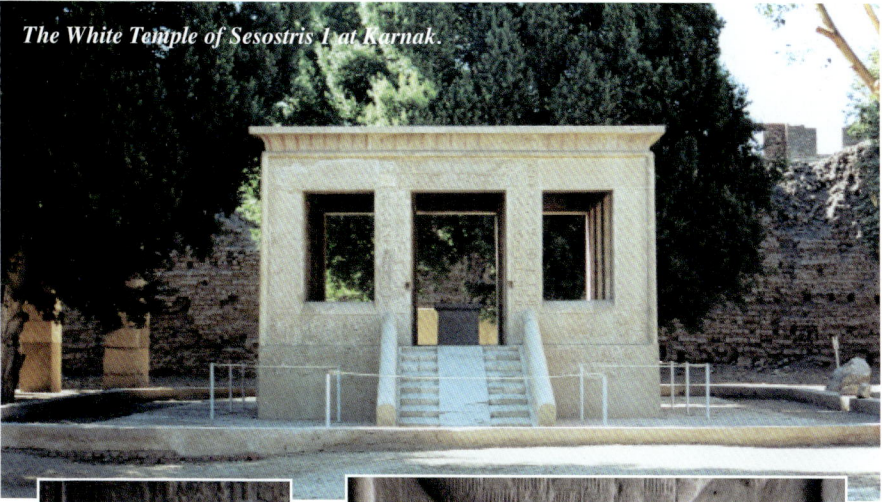

The White Temple of Sesostris I at Karnak.

Detail of the third column on the right showing the guineafowl used in the word eternity.

A close-up of the guineafowl showing the helmet and spots.

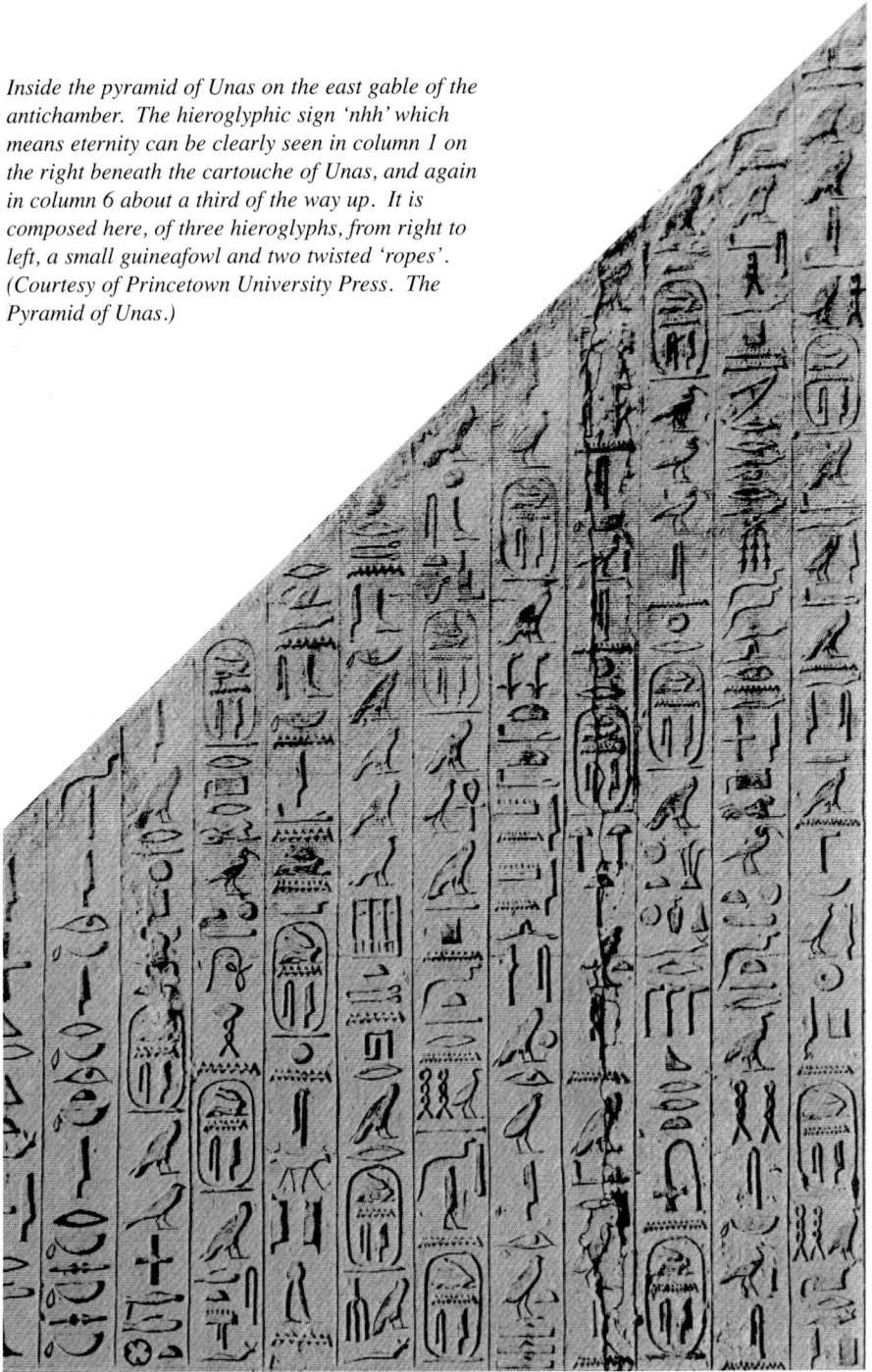

Inside the pyramid of Unas on the east gable of the antichamber. The hieroglyphic sign 'nhh' which means eternity can be clearly seen in column 1 on the right beneath the cartouche of Unas, and again in column 6 about a third of the way up. It is composed here, of three hieroglyphs, from right to left, a small guineafowl and two twisted 'ropes'. (Courtesy of Princetown University Press. The Pyramid of Unas.)

From the pyramid of Unas at Saqqara, showing the cartouche, second line up on the right. The hare is 'u', the wavy line is 'n', the feather is 'a' and the crook is 's'.

Above:

The entrance to the tomb of Senneferi, (TT99), Tombs of the Nobles at Luxor.

Left:

A painted hieroglyph of a guineafowl in Senneferi's tomb. (Courtesy of Nigel Strudwick, The British Museum.)

A greek vase or skyphos, dated 6th century B.C. (Courtesy of the Metropolitan Museum of Art, Rogers Fund 1941.)

A fresco of two guineafowl being hand fed, in the House of the Vettii, Pompeii, Italy. (Courtesy of the Ministry of Cultural Heritage and Environment, Italy.) 1st. Century A.D.

Above: The Viridarium, (green room or garden room) in the House of Centenario,
Pompeii, Italy. 1st. Century A.D.
Below:A very faded fresco of a guineafowl in the Viridarium. 1st. Century A.D.

Two guineafowl on a mosaic in the Terme di Diocleziano, Rome. This is in storeage and not on view to the public. It is just under a metre square. Note the guineafowl pecking the snails, bottom left. 4th century A.D. (Courtesy of Terme di Diocleziano, Rome.)

A mosaic of a guineafowl in the ruins of the House of Eustolios at Kourion near Limassol, Cyprus. There is an interesting inscription along the bottom of this mosaic which reads: "This house, in place of its ancient armament of walls and iron and bronze and steel, has now girt itself with the much venerated symbols of Christ." Surrounding the central guineafowl there are two fish, a pheasant and a magpie. 5th century A.D.

KEEPING CRESTED GUINEAFOWL
(GUTTERA EDOUARDI)
AND VULTURINE GUINEAFOWL
(ACRYLLIUM VULTURINUM)

The Crested Guineafowl is quite different from the domestic guineafowl, as it has a tuft or brush of small black curly feathers on its head. It is a quiet bird and can be mixed with other breeds in an aviary. It is found in the wild in South East Africa, Zambia, South Africa and Malawi.

Of all the species of guineafowl, the Vulturine is the most spectacular with its striking colours, and is the most sort-after for bird collections. It does not have the precociousness of domestic guineafowl, being a placid bird, easily hand reared and happy to be with other breeds in an aviary. It is found in the wild on the eastern side of Africa, north of Kenya and south of Ethiopia.

These birds are nearly always sold as breeding pairs; no one sells hatching eggs or chicks because of the potential value of the birds. Captive bred guineafowl are more manageable and less flighty, than imported birds, which never really settle down and are more susceptible to disease. Neither species is suitable for inbreeding.

An aviary for guineafowl should be as large as possible, minimum for one pair being 10' x 20' (3m x 6m) with a roof height of at least 8' (2.5metres). It is advisable to have a small mesh nylon netting roof so frightened birds don't scalp themselves, but this is not always possible if there are other things such as parrots or squirrels in the aviary as well. The accommodation should face south to get as much light as possible to stimulate the birds. The floor should be raised to keep it well drained, and should have a 6" - 8", (20 cms) covering of sand. Both species love to dig for worms, seeds and roots, but tend to suffer from foot and respiratory problems which can be caused by damp and cold conditions. At the highest end of the aviary there should be a heated sheltered area with Perspex roofing, where the birds can feed and perch, and shelter during cold, windy spells. Perches should be rounded 2" x 2", (5 x 5 cms) or 2" x 3", (5 x 7.5 cms) timber. The vegetation in the aviary should be sparse and tough as both species have strong feet and beaks. Provision for several nesting areas should be made by providing logs, bushes and tall grasses. If the aviary is large, screening vegetation is essential to give the pairs some privacy, even though two hens may share the same nest, a shallow scrape on the ground. Don't mix the two species in the same aviary, and avoid overcrowding.

Food for the adult birds must be varied as is their diet in the wild. Because they are so placid and can be hand fed, they are often put in aviaries with other birds such as parrots or cockatoos, and will help themselves to their food as well as their own. Their diet should include pheasant pellets, chopped vegetables such as lettuce, carrots, peas and beans, and several times a week a little raw minced meat or mealworms, etc. These birds are very susceptible to Gapes and other parasitic worms and require a regular worming programme with Flubenvet (Flubendazole). The life expectancy of these guineafowl in captivity is 15 to 20 years.

BREEDING:
Guineafowl lay one or two clutches of eggs each year with about 12 eggs in each clutch. Laying can start as early as March and continue until August. The creamy-buff eggs are always put in incubators or under hens where they take 28 days to hatch. It is interesting to note the thick shells on these eggs and the slight colour variations from hen to hen. Guineafowl actually make good parents in the wild but cannot be relied upon in an aviary.

The chicks are reared in a brooder with cloth or corrugated cardboard on the floor to prevent crooked toes and slipped leg or hip joints. It's very important to get their dietary requirements right as they need a high vegetable content of about 75%, chick crumbs 25% and protein 18%. Mix finely chopped lettuce with other vegetables such as carrots, peas or beans which have been put through the blender to make a rough mush. The chicks are slightly shy feeders to begin with but soon find that lettuce is a favourite food. Once they are past 12 to 14 days old there should be no losses.

This is only a very brief synopsis on keeping and breeding these two types of guineafowl. I am indebted to the Exmoor Zoological Gardens (Danny Reynolds) and Birdworld, Farnham, Surrey (Kerry Banks) for this information.

KEEPING GUINEAFOWL
ON A SMALL SCALE

There are four ways to purchase guineafowl:

A) Adults
B) Poults
C) Keets (day olds)
D) Hatching eggs

A) Adults. The only way to keep 'bought in' adults is in an aviary. If this is the first time you have kept guineafowl and you decide to put them in a pen for just a few days before letting them out, think again! They will wander off and probably settle down with neighbours or become fox fodder, so that would be a waste of money. Also beware of buying adult birds at auctions as you will have no idea of their age, and it is difficult, not only to tell their age but also their sex while they are in a cage. However, if you already have some guineafowl, new ones should settle easily into your existing flock.

B) Poults. These are in many ways similar to adults. Unless you have some mature guineafowls to put them with, they will start to wander as soon as their wing feathers grow, but normally older adult birds will take the youngsters on and they will all become one flock. This is quite a useful characteristic of the guineafowl.

C) Day old keets. This is probably the best way of buying guineafowl but they normally come unsexed so buy a minimum of 6 or 8 to ensure you have several females. They are best put under a broody hen so that they will imprint onto her. They can be reared under a lamp and will imprint onto humans and become very tame, but this really requires time and only 2 or 3 birds. If you rear 8 or more under a lamp they have no 'belonging' instinct, and later, when they are able to fly as young adults they will wander off. So under a hen is best with small numbers, and they will also learn to roost with their 'mum' in the hen house.

D) Not many people will sell you hatching eggs but if you come across some it is worth putting them under a broody hen. A large hen will take 14 eggs and a bantam 8 or 9 eggs. The incubation period is 28 days. One of the advantages of having guineafowl eggs under a broody is that you rarely get any broken or dented ones because the shells are so thick. Always check how fresh the eggs are because if an old one goes pop under the broody or in the incubator the whole hatch will be ruined.

Incubation of guineafowl eggs is slightly more difficult than hens' eggs because of the extra week involved, the thickness of the shell, and getting the humidity right during the hatching period. If you have had experience with incubators you should be able to cope, but if not then you may well be disappointed. Incubation is a science; a careful read of 'Incubation at Home', (Gold Cockerel Books) is recommended!

REARING ON A SMALL SCALE
Keets with a broody, in the garden or free range.
Do not disturb the broody when the chicks are hatching, and try to keep from peeping under her to see what you have, except perhaps once in the evening to remove a few egg shells. The hatching process can take up to 48 hours. Do not feed the broody during this time but allow her access to water. After 48 hours check to see how many have hatched and if you have any dead-in-shells. Remove the broody onto the short grass run of the broody coop to feed and drink. She will then make a very smelly mess which must be removed.

Feeding is very straightforward. Just use fresh chick crumbs for the first six weeks and move on to rearer crumbs or small pellets after that until week 10. I always feed the broody a small amount of wheat, about an eggcup full per day.

The chicks are timid for the first few days but then start running about looking for insects and becoming bolder. Once past the 5 day mark there should not be any deaths except from poor management. At this stage some of the young birds look a little like pheasant chicks, although the coloured guineafowl come in a variety of lavenders, fawns and white.

After a few days you will notice the wing feathers beginning to develop and the chicks starting to flap about their run. They don't like the damp or cold and are quite vulnerable to coccidiosis so always rear them on very well drained soil and short grass, or indoors if out of season.

After 3 to 4 weeks the broody can be let out of the run with her young. They will be slightly nervous but very curious about everything around them including anything flying over such as a crow or a hawk. At 7 to 8 weeks move them to a large house and they will roost on perches with their broody. You will notice that they have many chicken traits except for their appearance and calls of course!

The guineafowl young will always perch with their broody unless they are disturbed or become accidentally shut out one night. This will cause them to take to their wings and perch on the nearest tree or building. Once they have a taste for outdoor perching it is very difficult to get them back in their house at night.

Of course this can be overcome by pinioning as described on page 21.

If you are going to keep guineafowl free-range you will need plenty of space for them. When they are confined and crowded they will run up and down the fence making a muddy mess unless the pen is raised and has a sandy or well drained floor. If they are outdoors on grass, a well drained area at least the size of a tennis court will be sufficient for about 30 birds.

A little thought should be given to your neighbours, as guineafowl, although excellent alarm birds, are of course very noisy, so the siting of your aviary or outdoor pen will be all important. If they are allowed total free-range this should be less of a problem, only they do chatter to each other at night so you don't want them roosting too close to the bedroom window!

Guineafowl do not need much food except when the weather is cold. They live on vegetation and insects, and normally top up from the poultry feeders as and when they need.

Nesting is always on the ground, although having said that, I have seen guineafowl lay in nestboxes in a henhouse. If they are in an aviary they will need a choice of several well covered places, so use pine or fir branches to make 'wigwam' shelters for them.

When your birds are free-range it can be difficult to find their nests as they are usually well hidden in clumps of nettles or brambles, very inaccessible. When you do find them there is often a pile of eggs as several hens will use the same nest. Be careful not to trample down the surrounding vegetation too much as that would make the nest visible to both winged and ground vermin, and could possibly deter the guineafowl from coming back to use it.

When you have found the eggs the problem is which ones to take. If there are 30 or more, take about half and mark the rest with a pencil or permanent felt tip pen so you know which are the older ones next time you go to the nest. Of course you will have no idea if your eggs are fresh or partly incubated, but you can find out by candling them. See "Incubation at Home".

It is a good idea to put your guineafowl into an aviary if you can during the breeding season, as a sitting bird is very vulnerable to Mr. Fox, and it is also easier to control the egg collection like this.

Guineafowl make excellent parents and are very spirited fighters if there is a predator about. They will rarely sit on eggs in an aviary unless they feel secure and hidden away, although there are always exceptions to this rule. If you want your eggs hatched it is better to use a broody or an incubator.

If you wish to make pets of your guineafowl you must either hatch them yourself or buy them in as day-olds because they need to 'imprint' onto you from the very beginning. They can make good pets although rather messy as you can't house train them, and, depending on what they have been eating, they can be pretty smelly as well! I had one for several months called 'Peedoo' because of the call it made. They are very fidgety birds, always rummaging around in their plumage, taking off yet another piece of feather casing, and they do like to struggle up to your neck and sit on your shoulder to keep warm. They're not above preening your hair as well, in fact the hair in my ears became a fascinating but painful target!

HOUSING, BREEDING AND ARTIFICIAL INSEMINATION

HOUSING:

There are three main methods of housing guineafowl for breeding:
  A) Cages
  B) Raised open verandas with sheltered accomodation
  C) Free range netted pens with shelters

A) Cages   The cage system is really only for large commercial organisations who are breeding for improvement of blood lines and hatching eggs on a large scale. The cages are stacked 3 or 4 high, with 4 females per cage. Opposite in similar cages are the cock birds, one per cage. This arrangement helps to stimulate the cock birds even though they have no physical contact with the hens, and this helps the insemination process. It is most important to have the building and cages properly designed to facilitate automatic feeding and watering, egg collection (which normally uses a rollaway system), and the removal of dung either on a conveyor belt or into a collection pit. You will also need lighting and ventilation of course, and plenty of access to the birds for artificial insemination. Breeding in these conditions takes place throughout the year.

B) Raised open verandas   This system is used by the smaller commercial concerns. The birds have access to a wire mesh run or veranda so they get the benefit of the sun's natural stimulation. There are covered drinkers and

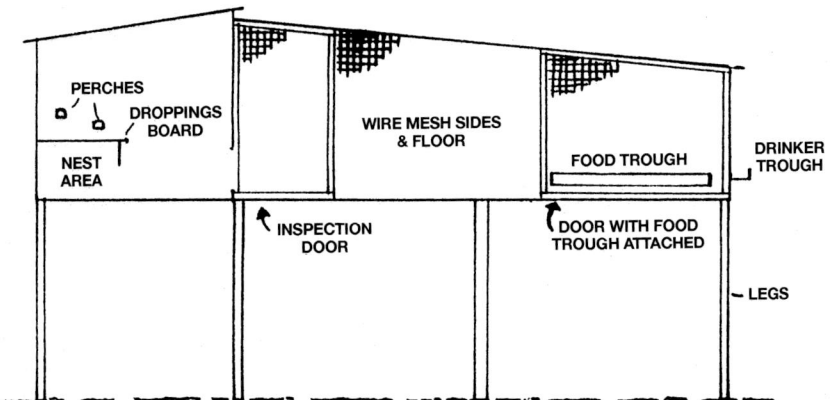

A design for a breeding veranda, 10' x 10' (3m x 3m). The height at the front of the veranda should be 3' (90 cms) and where it joins onto the house it should be 4' (120 cms). The house area is 10' (3m) long and 5' (1.5m) wide. The veranda roof should be made of clear perspex and wire netting. This system will house 30 females and 5 to 6 males.

16

feeders and a shelter or house for perching and laying eggs. The floor of the house is covered with straw, and some people put in a communal nest box although this is not always done as guineafowl all tend to lay in the same place. Perches are provided and droppings boards help to make the house cleaning easier. The mesh on the floor should be 3cms x 3cms. With this system you have 5 hens to 1 cockbird. This is an excellent method as it is very simple to run and very disease free.

C) Free range pens with shelters This is the cheapest method of housing guineafowl for breeding but you do need to be sited on well drained land. The system can be either semi-permanent or moveable, but the latter is best because fresh ground can be used each year. The size of pen is determined by the number of birds you have, but to give an example, 30 hens and 6 cocks would need a pen about the size of a tennis court. The birds don't need wing clipping as there is netting overhead, about 7 or 8 feet high, (2.5 metres) and the sides of the pen should be covered up to a height of 3 feet or one metre. This shelters the guineafowl from any cold cross winds. Place a few low rails inside the pen and stack some fir branches against them so that the birds can nest underneath. You can use automatic drinkers and a hopper system for feeding, and try to arrange a shelter with perches in the centre of the pen. Remember to position this run well away from your neighbours as the birds can be very noisy. If you put another run next door you can let your birds into it when the first one is used up and stale; don't forget, guineafowl are great grazers and diggers. Remember to open the door and let them make their own way into the new pen, if you try to drive them they will panic.

If you want to produce eggs all year round then a deep litter shed with controlled lighting and ventilation will work well attached to this system.

BREEDING:

Whether you are breeding indoors or outdoors there are three main criteria: the amount and duration of light, the ambient temperature and the correct level of protein in the food. It is easy to understand why the large commercial organisations run everything in controlled environments, so that if you visit one of these operations you will see very little, and what you do see will be behind glass because of hygiene precautions.

The three main criteria are as follows: light at 100 watts for a duration of 14 hours, ambient temperature 18 - 20 degrees C, and food 18% protein.

If you buy guineafowl from a large commercial organisation, you will find that, once they mature, they lack all the commercial traits of their parents. This is because they have been deliberately bred with planned recessive genes for sale to the public.

The commercial birds are some thirty years in advance of normal farmyard guineafowl and can reach a weight of seven and a half pounds (3.5 kilos). The genetics of these birds is complex. Needless to say, the geneticists are always pushing them to the limit in their efforts to produce a heavier bird that grows quickly. In the past the domestic species was crossed with some of the wild species to achieve this, but now it is done by line breeding. The problem is that the larger the bird gets, the less fertile it becomes and the fewer eggs it lays; therefore there is a constant juggling match going on, mixing the various breed lines in an effort to ensure speed of growth and good weight of bird, together with fertility, egg production, hatchability, viability and disease resistance. A quiet temperament is also important as the birds are handled once a week for insemination. If the mix of genes is incorrect there are problems with fertility, dead-in-shells, deformities, blindness, poor feathering and general viability. (Viability means healthy developement from chicks to keets to mature birds.) Add to all this the ability to produce colour sexing in chicks, and you can understand why these large companies guard their genetic know-how so strictly.

More and more people these days are wanting male and female chicks of a uniform size, but there is still a variation in egg sizes, due in part to the ages of the birds that lay them, and these eggs of course produce different size chicks. However, a breeding programme in France has, in the last few years, seen a levelling of the size differences between the two sexes. There is also a move to breed more dark legged birds instead of the more traditional ones with pink / orange legs. This is partly to do with sex-linkage and partly to differentiate the dead carcases of guineafowl and pheasants.

ARTIFICIAL INSEMINATION

Artificial insemination of guineafowl has been around for many years. It plays a vital role in research, preserving blood lines and improving stock, which random free range breeding could never achieve.

The hen birds start to mature at 12 to 16 weeks, but it is important not to push them to breed too early as this would result in small eggs and / or prolapses. So the young hens need to be gradually brought into breeding condition with the use of reduced lighting and lower levels of protein in the food. Young free range females eating high protein food will produce their first eggs as early as 12 weeks.

Weekly artificial insemination can begin at about week 28 when the females are ready for the males. It normally starts after lunch as the peak laying time is 12 noon. The staff usually wear gloves to protect themselves against the guineafowls' claws which are sometimes very long. A team of three people can inseminate 2,500 birds in a shift. The cock bird is taken from his cage and his abdomen is massaged to help arousal and open the cloaca. The penis pops out

A Crowned guineafowl. (Numida meleagris coronata)

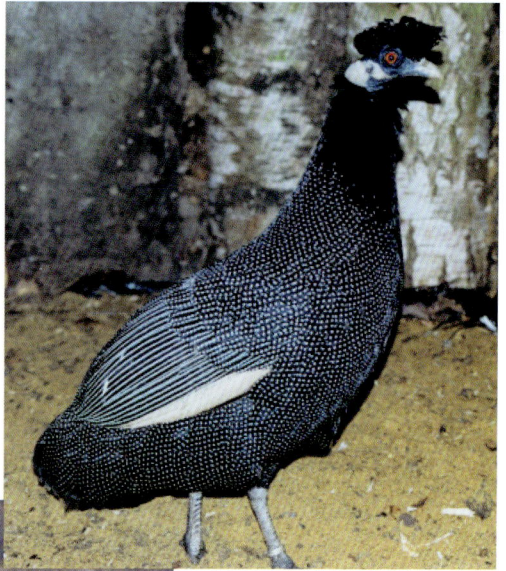

A Crested guineafowl.
(Guttera edouardi edouardi)
Birdworld.

A Vulturine guineafowl.
(Acryllium vulturinum)
Birdworld.

Day old guineafowl chicks of various colours.

A day old guineafowl chick showing the white egg tooth on the end of its beak.

One week old chicks showing the wing feathers coming through.

Two week old chicks, nearly able to fly.

Male or female? Normally the male has larger wattles but this is not always the case.

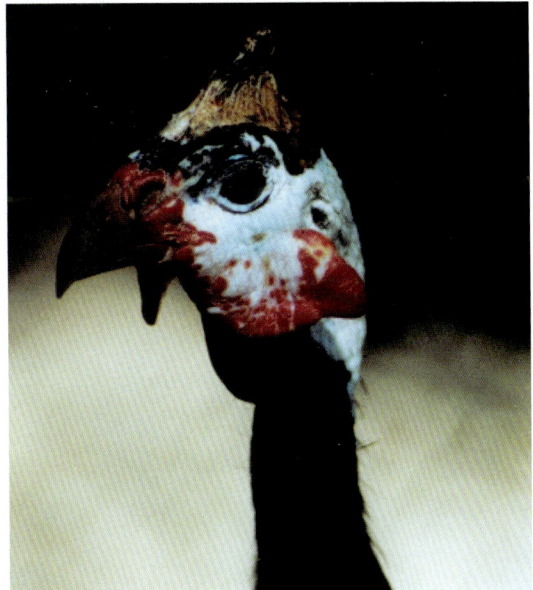

The white facial markings can vary from bird to bird. This picture clearly shows the guineafowl's horny crest.

Powder Blue

White African

Powder Blue

Porcelain

Lavender

Buff Dundotte

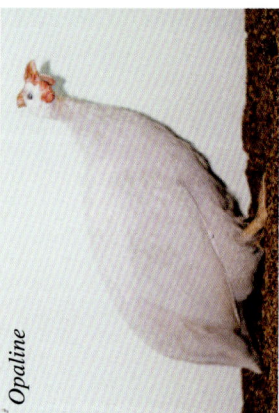

Opaline

Lite Blue

Royal Purple

Coral Blue

Pearl Grey Hen

Slate

Lite Lavender

Pewter

Violet

Chocolate

Brown

Day old chicks in a small brooder with a heat lamp.

Four week old keets. Note, the white and lavender birds are easily distinguished, and the darker ones are going through a brown plumage phase.

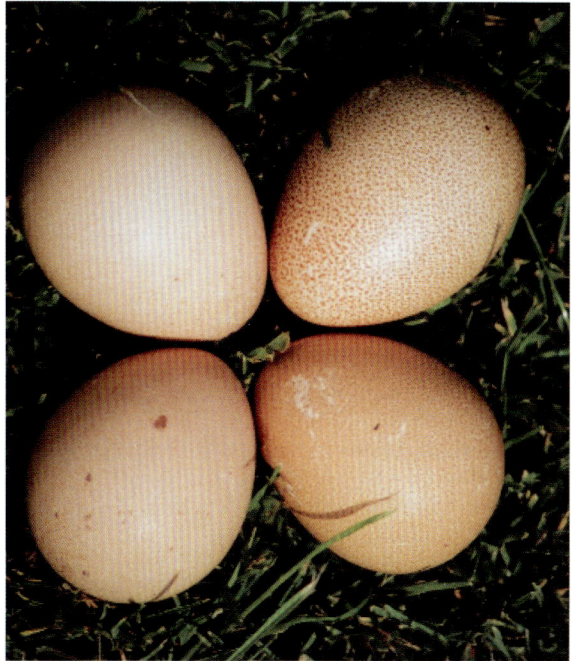

Guineafowl eggs showing the shape and variation in colours.

An oven ready guineafowl. Prices can vary enormously; this bird is very much at the top end of the market.

These are Snake's Head Fritillaries which I have included as their Latin name is Fritillaria meleagris.

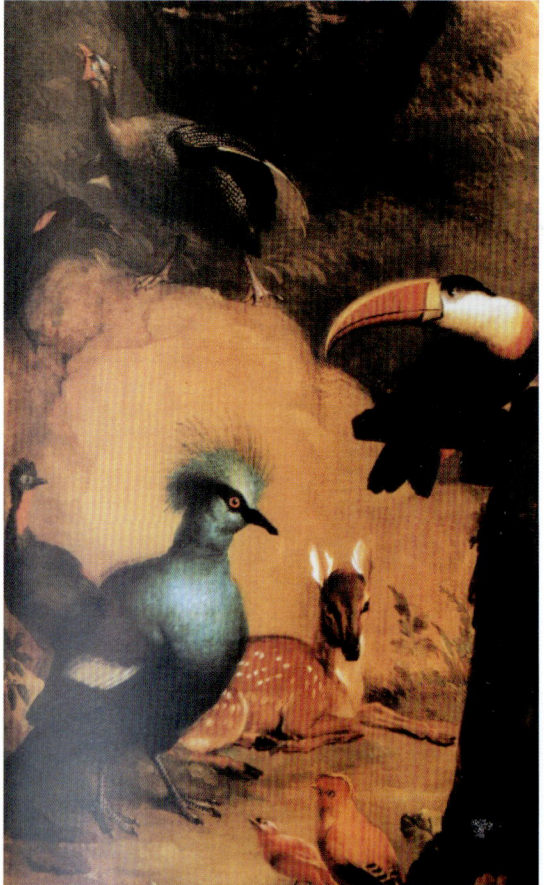

This painting by the Dutch master Aert Schouman, (1710 - 1792), includes a guineafowl on the huge canvas which measures 8'6" x 3'4" (260 cms x 101cms)

In Africa there is a range of crafts involving guineafowl, which use all kinds of materials including wood, coconut fibre and scrap metal.

Guineafowl art is also found printed on cloth.

and the sperm, a very small quantity of about 0.05 ml, is collected in a fine glass or plastic tube or straw. This is done by the operative who sucks on a plastic tube through a liquid trap so he doesn't get a mouthful. Normally two males are used for one straw (plastic tube) and then the sperm is diluted by 50%. This is necessary for several reasons: although the amount of sperm is very small it is actually extremely concentrated, and dilution means that not only can more female birds be inseminated but fewer cock birds are needed. Normally the sperm from two males would be sufficient for 14 to 16 females, but with dilution 28 females can be fertilised.

Time is of the essence when using this system as the sperm is only good for 20 minutes in the open, hence the neeed for quick and easy access to the cages, and plenty of room for the team to move around.

Once the dilution has been done the females are caught and massaged to help open the cloaca. Sometimes faeces are produced and will have to be wiped away. The straw of sperm is put into a special pump and offered up to the hen's cloaca where it is inserted into the vaginal orifice and a measured amount of diluted sperm is delivered inside the female. The birds become quite used to the routine after a while, and can be handled with ease by the operators. Of course the research area is constantly busy with records being kept, eggs marked and new batches of chicks being carefully monitored.

The laying period begins at week 30 to 31 and lasts for 36 weeks. During this time each bird lays about 180 eggs, the fertility is about 85% and hatchability roughly the same. These figures vary during the laying period as fertility is higher to start off with and gradually decreases over the following weeks.

If you are using a veranda or free range system, you might expect to collect about 160 to 170 eggs per female, but the fertility will actually be lower, between 70% to 75%.

Gallina Africana.

*From Willughby's
Ornithology, 1678.*

EGGS, INCUBATION, HATCHING, COLOUR BREEDING AND PINIONING

EGGS

Guineafowl eggs vary in size and colour, but, as with quail, you can often identify a bird that laid a particular egg by the colour of the shell. There are various shades from light brown to biscuit to off-white, with some eggs showing brown spots or freckles. Double yolkers are recorded mainly from older birds. The shell is very thick when compared with a hen's egg, 15% in a guinea and 11% in a chicken as a percentage of the total mass. A thick shell has its advantages: for instance there are far fewer breakages under a broody hen as these eggs rarely 'star' or dent and can take a certain amount of rough handling. The Russians have experimented with guineafowl eggs on earth and in space. They frequently sell them in shops because the tough shells make them easier to handle, and they are also more likely to survive the sometimes crude methods of Russian transport.

The eggs are nearly always very conical at one end, although this is changing somewhat due to selection and intensive breeding. One always connects conical eggs with sea birds, as this distinctive shape prevents their eggs from rolling off cliff ledges. It is also a characteristic of turkeys, peahens, plovers and various other ground-nesting birds: they all have conical shaped eggs which pack more neatly beneath the hens, taking up less room than more oval shaped eggs.

The length of egg varies from 1 7/8" to 2 1/8", (4.8 - 5.4 cms). Some of the difference is seen during the laying period when the first ones are quite small. Of course the weight varies as well, from 40 - 59 grams. This is very awkward for commercial producers who like chicks of a uniform size, and most large hatcheries will not grade their eggs prior to incubation. The yolk size is about 33% of the total egg weight.

Guineafowl tend to lay at lunch time, rising to a peak at about 2 pm, so egg collection should be round about 4pm. If they are to be used for incubation you should carefully weed out any misshapen specimens or ones that are too large or too small. Some people recommend washing eggs, but I believe that there is a reduction in the hatch when this is done. If the area where the birds lay is kept clean, and they normally just use one spot, there should be no need to wash the eggs. Other people have different ideas however, and if necessary use a special preparation in warm water to clean them with.

Allow the eggs to settle for two days in a store room at an even temperature of 18 to 20 degrees C before incubation. You can keep them for up to 7 days for the best hatching results, but you will need to turn them daily. The large commercial hatcheries turn or tilt their eggs every two hours.

INCUBATION

The incubation period for guineafowl is 28 days, but this can vary by up to two days with small incubators if the temperature is incorrectly set. The correct temperature for incubation is 37.5 degrees C with humidity running at 65% until day 25. From then until day 28 humidity rises to 80% with a slight decrease in temperature of one degree. Candling is important at day 10 and later at day 20 so that you can check developements and remove any infertile or dead-in-shell eggs. For further information see our book "Incubation at Home".

HATCHING

Large commercial hatcheries claim that they get a 95% hatch. Small scale hatches tend to be nearer 70% to 75%.

The chicks weigh approximately 32 to 35 grams. At this point you should remove any deformed, splay-legged or non-viable ones before moving the others to a rearing pen or boxing them up for a client.

COLOUR BREEDING

As you can see from the photographs in the centre of the book there are already over 18 different colours of guineafowl, some more striking and more easily defined than others.

Originally there were three basic colours, white, pearl and lavender. Other colours have been developed by crossing Domestic with Helmeted guineafowl, and if you look carefully at them you will see that some carry many spots in the feather, some just a few and others, such as the white guineafowl are a clear colour with no spots at all. It is interesting to note that the first plumage of all young guineafowl except the white, lavender and pied birds, is the natural brown colour, and after the first moult at 8 to 10 weeks the true colours start to appear.

PINIONING

This operation is legal in France but I am unsure of the legal position in this country.

If the chicks are destined for a free range rearing unit, the client may require them to be pinioned. The outer portion which contains the flight or primary feathers is cut from one wing only. There is a special double-bladed knife available for doing this which cauterises the wound at the same time. In fact there is hardly a drop of blood and if you do it this way the wound is sealed from infection. For small scale operations a pair of sharp nail scissors works well. If the birds are sexed some people like to cut one wing for males and the other for females.

SEXING

Day-old sexing is done by professionals, but they do require large numbers to make it economically viable. If you know a local one that could be very handy. They normally work in teams and have a high percentage of accuracy.

They pick up a chick and open the vent to reveal the cloaca, having first removed a pellet of dung. They study the shape of the cloaca through a magnifying glass or special spectacles and this will tell an experienced person the sex of the chick. Strong lighting is vital for this job.

Sexing is important for the large hatcheries when they are choosing breeding stock, and also for the capon trade, (see page 30).

There are five different ways of sexing mature domestic guineafowl.

A) Noise or cry. The female bird has a repetitive two syllable call, sometimes described as 'come back, come back' or 'buckwheat' or 'swap hat'. The male call is just a single syllable screech which is normally used as an alarm call. The best way of sexing free range birds is to listen to their cries, but you will have to wait until they are 10 to 12 weeks old. Once you have identified a bird by its call you will have to keep your eye firmly on it until it is caught. It is possible to colour code your stock with leg rings, but take care not to harass your birds when you catch them as the females will go quiet and not utter their distinctive cries!

B) Observation. It is said that male and female guineafowls can be identified by the size of their wattles. To a certain extent this is true, but as they get older females will become more male-looking, and in certain blood lines the male and female look very similar. You will notice a difference between the two during the breeding season. The male displays with a jaunty walk, arched wings and often a single syllable call, 'tut, tut'. The female on the other hand appears to carry her body lower to the ground and takes little notice of the cock bird tripping round her.

C) Colour. There are sex-linked guineafowl which are bred 'in-house' for the commercial producers. These are not available on the open market. The colours used are a very dark male similar to the Royal Purple and a light buff / yellow, spotted female.

D) Physically and visually. If the female is 'in lay' her pelvic bones will be one to two fingers width apart, but otherwise they will give no indication of sex. Birds that are used to being handled can be physically sexed by experienced people. The reproductive area is gently squeezed and massaged with a thumb and forefinger. In the case of a male the organ will pop out of the circular cloacal opening, which remains rounded and less pronounced in the case of a female. Some male organs are far more easily exposed than others, so mistakes can be made by the inexperienced.

E) Surgically. This is a method favoured by breeders of wild guineafowl such as Crested and Vulturine, and is only practised by professional sexers.

A guineafowl from Ulisse Aldrovandi, 1660.

COMMERCIAL REARING,
INTENSIVE AND FREE RANGE

The first 4 weeks of rearing guineafowl commercially are the same whether they remain indoors or go out on free range later. After this period the systems go their seperate ways.

Intensive indoor birds are kept for about 2 months and their weight at the end of this period is 1.4 kilos or 3 lbs. approximately. The free range birds are kept for about 3 months and reach a weight of 1.9 kilos or 4 lbs. approximately.

PREPARATION However large or small the batch of birds you are going to rear, you must be properly prepared, so we will look at a check list of things to do before the birds arrive. It is important to have everything up and running three days beforehand so that the house has time to dry out and come up to the right temperature, 27 degrees C, and you can make sure everything is working as well.

HOUSING The housing used for guineafowl chicks is the same as for broiler chicks. Two important factors are insulation and ventilation. The correct level of insulation will save you a fortune in heating bills; take advice about this as it is a specialist subject. Ventilation comes in two forms, natural and fan assisted. Natural ventilation is created by the design of the building which should allow the warm, stale air to drift out of the top through the roof vents without causing any draughts inside the house which would affect the birds. Fan assisted ventilation is used in addition to natural ventilation and can be thermo-electrically controlled to produce an even temperature and help eliminate dust and smells. Of course the house must be thoroughly cleaned and disinfected and allowed to dry before use. It must also be checked to make sure it is rat, mice and wild bird proof.

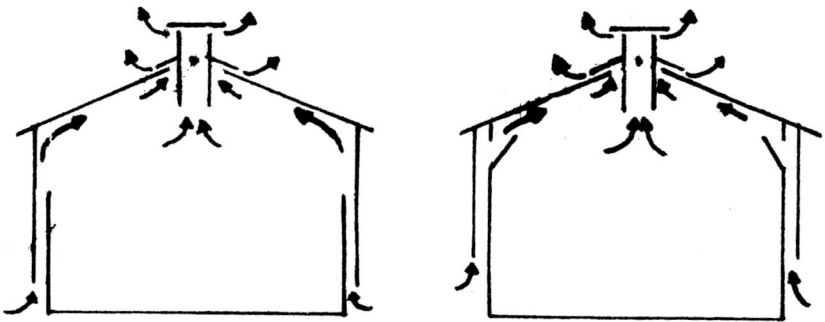

Two types of housing construction showing two different methods of natural ventilation which incorporate fan assisted ventilation as well.

HEATING Most farmers use Calor Gas either in tanks or cylinders, because of ease of control and independance of supply. It is cheaper as well! Power cuts always come at the most crucial times, and on-farm generators can usually cope with the lighting requirements, but not the power needed for the heating elements. The heater should be placed above the chicks, giving a floor temperature of 36 to 38 degrees C and a room temperature of 27 degrees C. These temperatures are maintained for the first week and then the heat is gradually turned down as the birds feather up, until it gets to 23 degrees C. This is controlled by the ventilation system in conjunction with the heaters. There is a certain amount of body heat given off by the birds, and of course the temperature outside will also be a factor, so the electric controls are constantly monitoring the prevailing conditions to maintain the temperature that has been set on the thermostat. It is always advisable to have several thermometers placed round the inside of the building so you can check them if the house feels too hot or too cold. The ventilation must be able to cope with the increasing amount of ammonia produced from the birds' droppings, particularly towards the end of the rearing period.

LITTER Wheat straw is the best litter as it is not as dusty as shavings and cannot be so easily ingested by the keets. They are far more skittish than chickens, specially during weeks 3 to 5 while they are growing their first feathers, and will run about all over the place, a sure sign of good health. The wheat straw is disturbed far less by all this activity than shavings would be, as it mats down flat on the floor, and consequently there is not so much dust and fewer breathing problems. The straw must be clean and fresh, not mouldy at all, and should be 6" to 8" deep, (15 to 20 cms).

FOOD This needs to be ordered and in stock about 4 days before the chicks arrive, and supplies maintained during the rearing period. I was interested to see on a French farm we visited, that the chick food was slightly scented to attract the birds to it. The mixture and balance of the feed is a complicated affair, very often dictated by world prices of the components of the diet. You may sometimes find that a new batch of food will arrive on the farm from the manufacturer you have always used, and yet when you open the bag the contents look and smell completely different from the last lot, although the dietary balance is the same. It is probably better to pay slightly more for your feed and ensure continuity, and also use a producer who knows what guineafowl require for the best results. The protein content in the food is 18% and most manufacturers use a vegetable protein such as soya.

WOODEN SURROUNDS FOR CHICKS It is vital to keep your chicks inside a safe area for the first week to ten days. Guineafowl chicks have a habit of wandering off away from the heat and food and water, so the surrounds keep them together and prevent this. Some people use straw bales but I have found that the chicks can get themselves wedged in the crevices, so a seamless wooden surround is best. A sheet of hardboard 8' x 4' (244cms x 122cms) cut in half lengthways works well. As they grow so the surrounds are extended to make a bigger area. It is easy to tell with this system whether the birds are too hot or too cold by the way they are spread out or crowded together. The surrounds can be taken away after 10 days. The corners of the house must be blocked off with a wooden or wire mesh barrier to prevent smothering.

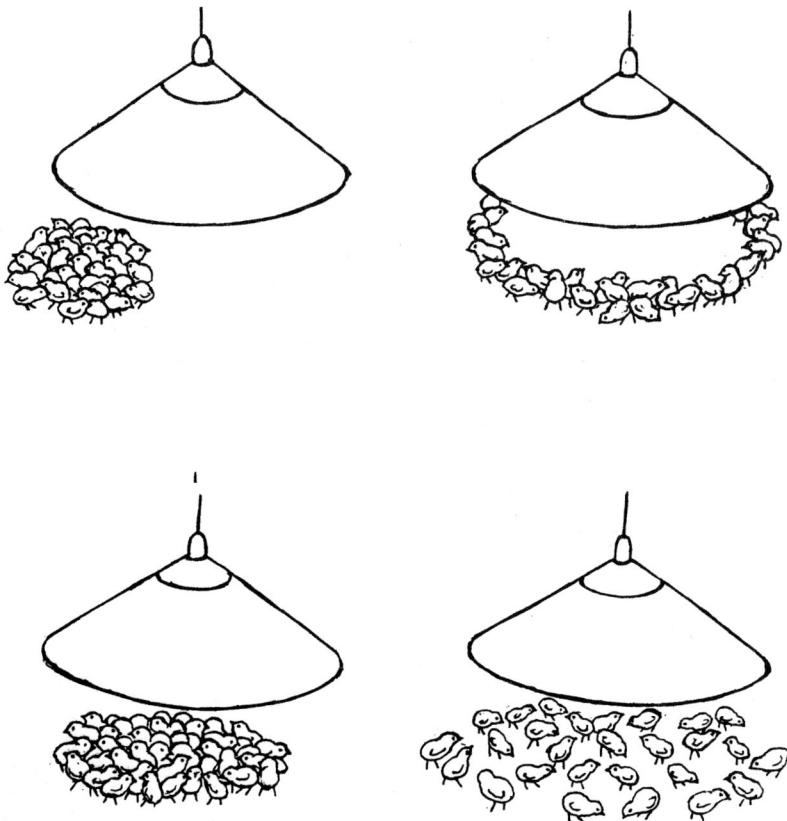

These diagrams show how chicks disperse or crowd together if there is a draught in the rearing area, or if they are too hot or too cold under the heat lamp. It also shows their distribution under the lamp when all conditions are ideal.

26

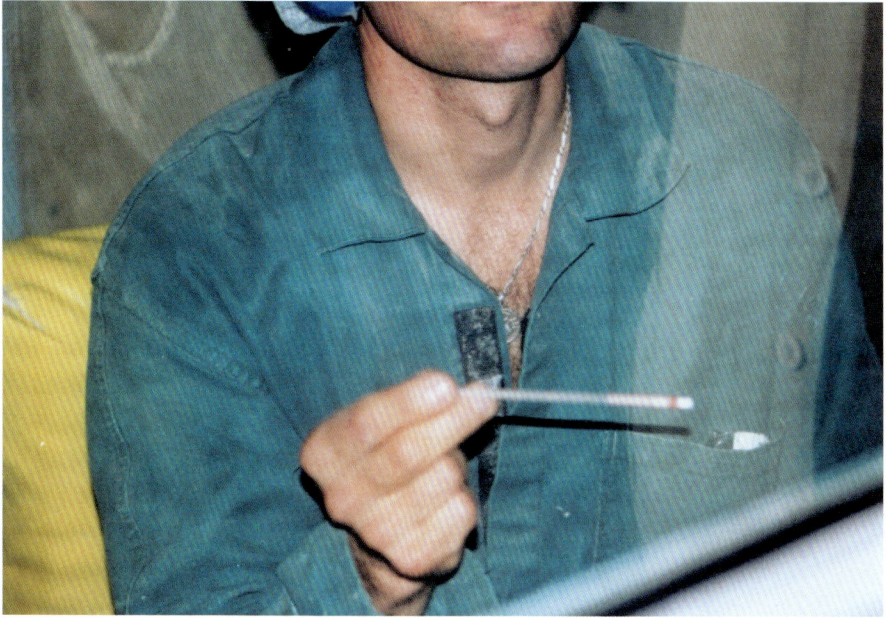

A straw of guineafowl semen prior to being used for artificial insemination. (Galor)

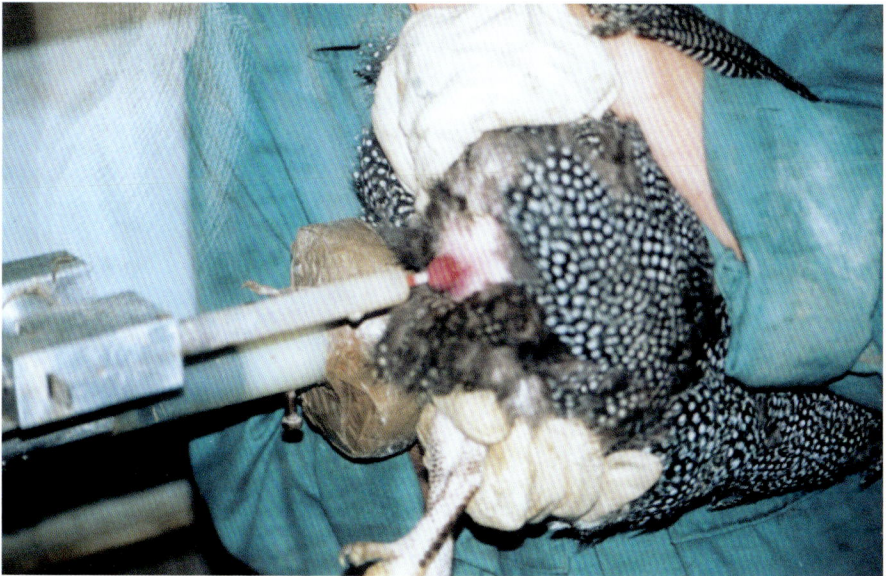

A female guineafowl being inseminated. (Galor)

Eggs being stored before going to the incubator. Note the piston for turning or rocking the eggs, timed for every four hours. (Galor)

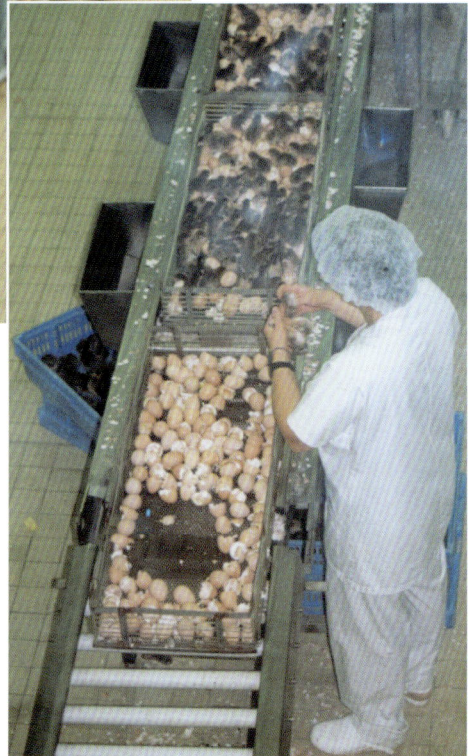

After hatching, the chicks are removed from the hatching trays and sorted. This operator is looking for leg deformities or distended navels. (Galor)

The good chicks are put into the chute beside the operator, and end up in a plastic box, ready to be counted and boxed for transport. (Galor)

A good view of nipple drinkers with an electric line along the top to stop the birds from perching on them. Note the straw bedding and egg trays used as feeders.

The young birds have already found the food in the automatic feeders, and are scratching around in it. The feeders will have to be raised slightly from the floor to prevent food wastage.

This is a 28 day old bird with its first brown plumage, and the second grey plumage coming through on the neck and shoulders.

The author (right) with Jean Olliver (centre) of Savel, outside a rearing house with the local farmer. Note the health precautions we had to take before entering this building.

A healthy looking batch of 28 day old birds in an intensive system.

Free range guineafowl about 12 weeks old and nearly ready to go. These birds looked magnificent, running round the orchard in noisy grey flocks.

This is the rearing house where the birds spend the night. Note the long, low sliding door for access.

This is a Light Brahma / Guineafowl cross that occured at the Domestic Fowl Trust in 1990. It lived for 5 years.

Hmph!... Crested is she!

A postcard from Zimbabwe, (Donald King) courtesy of naturals @ icon. co. zw

FEEDERS Most people use automatic feeders connected to a bulk tank at one end of the building. These must all be disconnected, thoroughly cleaned and allowed to dry before assembly, and all working parts and bearings should be inspected and checked. For the first week to ten days, cardboard egg trays are used on the floor along with the hoppers, to help the chicks to feed. The hoppers should also start off on the floor, but remember not to put them or the egg trays under the heat source as it will affect the vitamins in the food. Use one hopper for about 25 to 30 birds.

DRINKERS These can be the cause of disease problems, not because they haven't been cleaned properly, but because of leaky valves and connections. There are four main types used in guineafowl production: float drinkers, fountain drinkers, cup drinkers and nipple drinkers. Float drinkers and fountain drinkers are used to begin with so that the guineafowl chicks can find water easily and get off to a good start. Float drinkers can be a real nuisance though, as they sometimes leak if the float gets jammed or stuck, but they are removed along with the fountain drinkers after 10 to 15 days. Cup drinkers have a small valve and are self-filling to a predetermined level, and nipple drinkers require the bird to peck the release valve to obtain a droplet of water. The drinkers start off on the floor and are gradually raised up as the chicks grow, to keep the water free of litter and faeces, and of course they are always kept away from the heat source. Use one cup or nipple drinker for about 20 to 25 birds. If you have a smaller scale operation it should be possible to dip the chicks' beaks in the fountain drinker when they arrive as this is an excellent tonic, but if you are dealing with thousands of them it is obviously not possible, so you hope that they will find the water by stumbling against a drinker or being attracted to its red base.

LIGHTING This is nearly always provided by electric bulbs in conjunction with dimmer and timing switches. To begin with the chicks are given 100 watt lighting 24 hours a day for the first 7 days. This is because they must see to find their food and water in order to survive. After the first week the light is gradually reduced down to 25 watts at three weeks. The lighting is never normally turned off until the birds are nearly grown up and then only for an hour, so they are reared in perpetual twilight to keep them calm. Some farmers do turn the lights out for one hour in twentyfour when the birds are feathered up as this allows them to get used to power cuts and the time it takes the farmer to switch on an alternative source. If there is no lighting or the power cut goes on for too long, there is a real chance of the birds becoming disorientated and smothering in corners, causing enormous losses.

INTENSIVE SYSTEM The birds remain in the house and there are several checks that must be made before the rearing programme can continue. Feeders and drinkers need to be raised so that the birds do not foul the food or water. Most people use an electric wire along the top of a row of feeders and drinkers to stop the birds perching on the supply line. You must check for water leaks continuously which means you must walk up the line at least twice a day.

When you are in the house, putting down fresh straw for instance, always move slowly so you don't frighten the birds. The heating and ventilation need to be constantly adjusted as they grow, produce more body feathers and make more manure. Always keep the lighting low so that they stay calm and don't start scrapping with one another. Sometimes birds are ruined by scratch marks and cuts, and they can kill each other if they fight.

If your birds are to go into laying cages, you should place some weld mesh grids on the litter two weeks before the transfer so that they can get used to standing on wire floors.

The stocking rate is 15 birds per square metre, which is a little over the size of a sheet of A4 paper per bird, and the mortality percentage is about 3%.

FREE RANGE At 4 to 5 weeks, depending on the weather, the low sliding doors on the rearing house can be opened. Only a small area should be made available to start with so the young birds can learn to go in and out and can become acclimatised to the open air. All the feeding and drinking continues inside as normal throughout the rearing cycle, because the birds need to feed and drink in the early mornings before they are let out. This also helps to stop wild birds from raiding the food hoppers, as they are usually reluctant to enter the house. Once the young guineafowl are 6 or 7 weeks old and are more confident in the outside world, you can put some automatic drinkers and hoppers of whole wheat out on the range for them.

The range should be as large as possible and the birds must be provided with shelter from the elements. An old orchard is ideal providing your birds are wing clipped. If you haven't any trees you can make a shelter with some sheets of corrugated iron raised off the ground about 30 inches (75 cms), on a wooden frame. Guineafowl are great foragers, but make sure you keep the grass short in their enclosure. They will peck at fruit such as apples and plums, and plenty of mineral grit must always be available for them.

One of the main problems with rearing free range guineafowl in this country is the weather, and because of this you should put down some hard standing for 5 to 10 yards (5 to 10 metres) around the house to avoid muddy conditions outside the doors. This problem hardly exists in France because of the warmer and drier climate. Another problem in this country is foxes, so electric fencing is a must to stop any incursions.

Free range guineafowl need to be fenced. In France a simple 5 foot (one and a half metre) wire netting fence is used because the birds are pinioned. It is doubtful that this operation is allowed in this country, but if the French, who are subject to E.U. rules and regulations, can do it , I cannot see why we shouldn't be able to as well! No doubt I shall receive letters from all sorts of National organisations telling me to retract this last sentence, but I am not sure how the Law stands if day-old chicks arrive in this country from France with their wings already pinioned.

Brailing the birds is not an option. This is where one wing is constrained by a cotton or leather tape, and as the birds grow the tapes have to be adjusted several times.

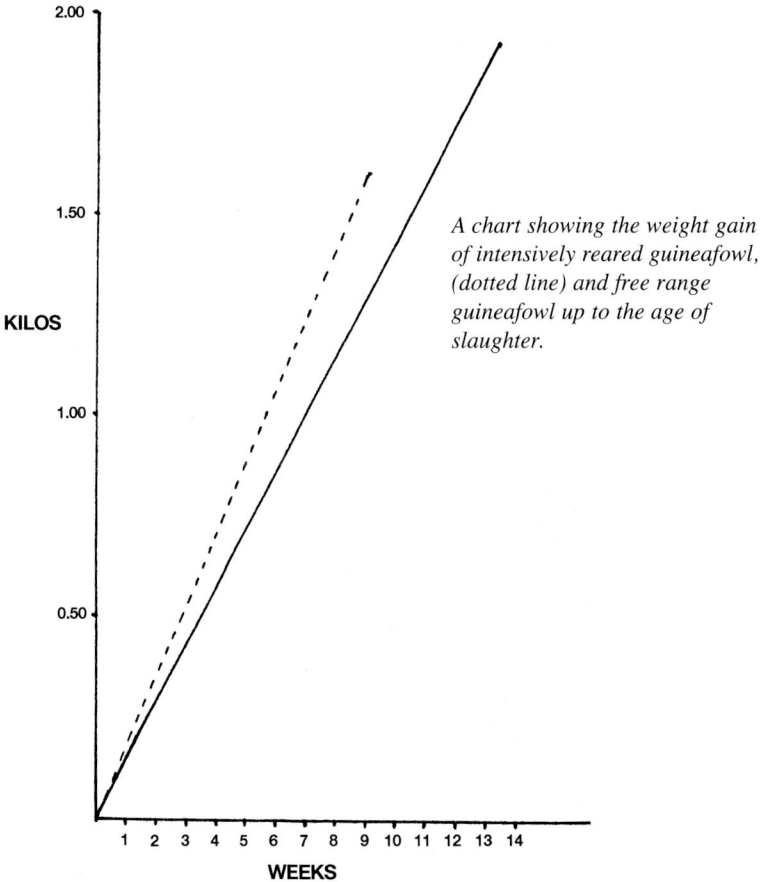

A chart showing the weight gain of intensively reared guineafowl, (dotted line) and free range guineafowl up to the age of slaughter.

Presuming that the birds are full wing, then the rearing area for a small scale operation would have to be netted. This does involve extra cost and labour but can be done quite simply.

Looking at both systems of rearing in France, the intensive and the free range, we found there was no comparison between the two. The mortality rate was lower in the free range system and the birds looked much better for being out of doors. But you have to weigh up the advantages of both, bearing in mind that one has a rearing period of two months and the other a rearing period of 3 months.

CAPONS This is a new market which has been developed by the French, a product to rival the turkey particularly at Christmas and the New Year. The birds are surgically castrated and then grown on for 6 months, attaining a weight of about 6lbs (two and a half to three kilos). The number produced in France in 2001 was about 80,000, but the French believe this number will expand rapidly. The old method of injecting a hormone pellet has been discontinued as there were problems of hormonal residue in the carcases.

CATCHING, KILLING, PLUCKING, EVISCERATION AND TRUSSING

You must withdraw the food at mid-day before the night that you plan to catch your birds prior to sending them to the abbatoir. Normally the hoppers are winched into the roof space. At night-fall the water is turned off and the drinkers removed which will give plenty of space in the house for the catchers.

Catching always takes place at night when the birds are sleepy and quiet. It is a dirty and dusty job and may require up to 20 people to clear a large house. The method used is to drive a certain number, a couple of hundred birds for instance, into a catching area, see diagram. There are three main operations involved: catching the birds carefully by the legs holding several in each hand, counting and loading them into crates, and carrying the full crates to the vehicle and empty ones back into the shed. The number of birds in a crate will vary according to the size of container, but once a figure is established, everyone will know and keep to that number. The crates are placed three high so that they are waist height for loading the birds. It is important to have an economically viable number in a crate, but not too many as overcrowding will cause them to sweat and die of stress. The lighting in the house must be as dim as possible and the ventilation will need to be turned up to help remove the dust.

This whole process has to be co-ordinated with the abbatoir, who normally kill the birds early in the morning. The time between catching and killing must be as short as possible. Of course careful planning is required from the point of view of timing, transport, number of crates, personnel and deliveries of foodstuffs.

Commercial killing is very much a numbers game, getting as many birds as possible through in an hour. The birds arrive in their crates, are unloaded and hung up on special metal brackets by their legs. A conveyor system takes them forward to be electrically stunned, and then they are cut and left to bleed, before being plunged into scalding water and fed into the plucking machine. The next process involves the insides being taken out by an eviscerating machine which cuts into the vent area and sucks out all the contents of the body cavity. Then the neck, head and legs are removed and the bird moves forward to the next section for finishing, preparation and packing. Some birds are packed individually, some are bulk packed and some are halved or quartered according to the demands of the market.

Today there are so many rules and regulations about the processes in abbatoirs that they would fill a book, so it is best to visit one of the smaller ones to see what is involved. It would probably be best to have your birds processed for you to begin with, until you see your way forward to either building a small abbatoir or sharing one, but any decisions of that kind would have to be carefully considered along with all other aspects of marketing.

If you have only a few birds to deal with for home consumption, you will find the following system works well.

Preparation for killing:

Starving: the birds must be starved for 12 hours before they are killed to ensure that the gut is empty. This prevents degeneration and discolouring of the carcases. Withhold grain for a 12 hour period before that as it takes longer to get through the system. The birds must have access to water.

Catching: the best way to catch your birds is to run them into a building they are familiar with and close the door. If you are planning to kill them the next morning, catch them at night and put them in crates, but not too tightly packed as they will sweat. Otherwise make a 'race' and catching area as per diagram, and take them out one by one.

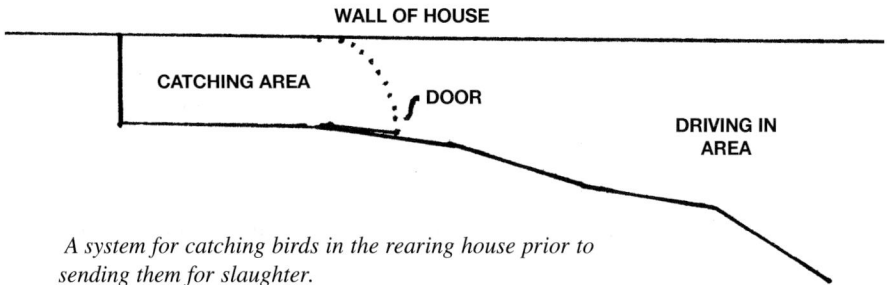

WALL OF HOUSE

CATCHING AREA

DOOR

DRIVING IN
AREA

A system for catching birds in the rearing house prior to sending them for slaughter.

Killing: humane dispatcher: this is a device for severing the spinal column without breaking the skin. It has the advantage of being instantaneous and the job is done quickly and cleanly. It also makes a gap for the blood to drain into.

Plucking: hand plucking is no longer an option today but it does give a good finish.

Machine plucking, two sorts:

1) Dry plucking: this machine has revolving plates or wheels to which the bird is held and slowly rotated. A fan sucks the feathers up and away into a bag. These machines do a good job, speeding up the whole operation to about 5 minutes per bird, although the tail and wing feathers still have to be done by hand. The machines can also be worked by unskilled people.

2) Wet plucking: this is more complicated as the bird has to be semi-scalded. The water has to be 125 - 130 degrees F; if it is colder the feathers will not be lifted, if it is hotter the skin will be reddened thus downgrading the

carcase. A thermostatic urn holding about 20 gallons will be the right size. The bird should be in the water for about 20 seconds, and is then held against rotating rubber fingers which take off the feathers. It must then be gutted immediately otherwise the skin will become discoloured. The operator must always take care not to get scalded.

Cooling: it is important that the birds are cooled down quickly to about 32 degrees F to inhibit the growth of bacteria. They should not freeze but the heat should be removed from inside the carcases, bringing the temperature down to 35 dgrees F. The birds should be hung in a fly-free, temperature controlled area with enough space around them for the air to circulate freely. A stone or brick building which is suitably painted, vermin proof and clean will probably be cool enough in an English winter, but beware of warm spells which could raise the temperature high enough to cause rapid degeneration of the carcases. Air conditioned cooling rooms are the norm today.

Evisceration or gutting: it is vital that conditions are kept as hygienic as possible. Do not use wooden surfaces as these cannot be cleaned properly. Melamine is acceptable, stainless steel is best.

Start by cutting the skin on the front of the shanks to expose the tendons. These can be removed to tenderise the legs, by looping them over a hook and pulling. Cut through the hock joint leaving a flap of skin on the back of the joint to prevent the skin shrinking up the leg during cooking.

Cut round the neck near the head and then cut the head off. Cut the skin of the neck down the back to between the shoulder blades using heavy shears. The gullet, crop and windpipe can then be carefully removed by cutting as close to the body cavity as possible. This leaves a flap of skin which is folded under the bird to keep the juices in when cooking and provides a space for the stuffing if needed. Next cut off the tips of the wings. Make a cut between the vent and the parson's nose, push your finger into the cavity and loop the intestine round it. Cut round the vent leaving it attached to the end of the intestine, then enlarge the hole so that you can get two fingers into the body cavity. You will feel the gizzard about the size of a small egg; pull on it carefully and most of the organs should come out. Take care not to break the gall bladder which is green, as it will taint everything. Seperate the heart, the liver and the gizzard, the lining of which is easily removed after cutting it lengthways. All these constitute the giblets. Make sure the body cavity is empty before washing it out with cold water. Wash the outside of the bird as well then hang it up to drip. Wash the giblets and wrap them in cellophane. Pat the bird dry with kitchen paper or a disposable cloth, then put the giblets inside the body cavity.

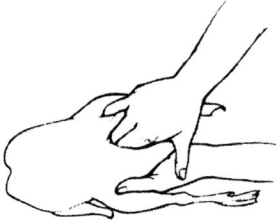

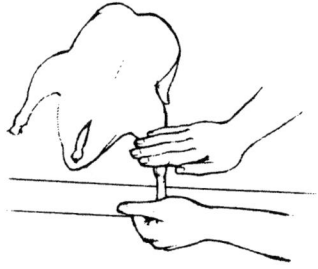

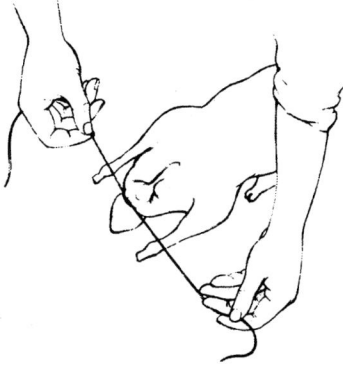

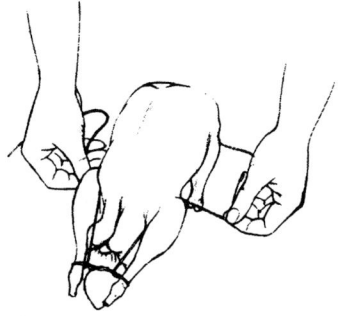

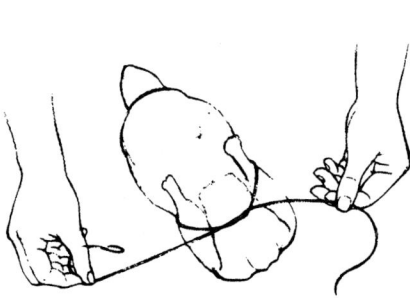

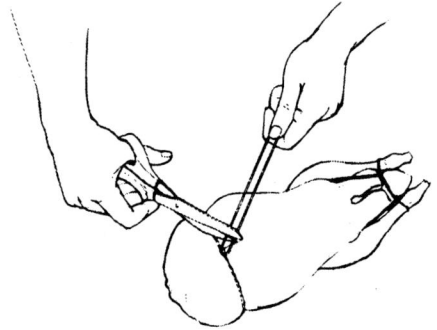

Trussing:
this happens to be a duck but the principals are the same.

34

PACKING AND LABELLING

Of course it is mandatory to label your product with your name, address, the sell by date and weight etc. It is equally important to have a label which gives the customer confidence and is clear and informative, telling them how the birds were reared and what they were fed on for instance. Remember, do not use more than 27 words as people do not have time to read more than that. It is probably best to get a suitable label designed with outside help. See what other people do, create your own logo, and make sure that your labels are not too big or too small, and that they conform with any existing rules and regulations.

Guineafowl are usually packed whole in plastic bags with a tied seal, but sometimes come in portions packed on trays.

The rules on Health and Hygiene seem to change quite regularly so it is best to contact the relevant agencies such as the Trading Standards Office and Health and Safety who will no doubt shower you with leaflets. These Government leaflets can be quite daunting, but a visit from a local officer of a particular agency can be very helpful.

It might also be a good idea to visit, if you can, a friend or company in your line of business, and learn from them of their experiences when dealing with these agencies.

You may find there is a possibility of forming a co-operative by sharing facilities with some other producers. One person could supply pheasants, another chicken and another duck, with perhaps some turkeys for the Christmas market as well. This would make sense, as premises for slaughter and preparation are expensive to set up. If you are already supplying pheasants you could add guineafowl to extend your sales season. There are many good reasons for going into guineafowl, but don't be too ambitious, at least to begin with!

I have only covered meat in this section but guineafowl eggs are also very marketable. I well remember a local hotel coming up with a novel idea for Sunday breakfasts - three different boiled eggs, hen, turkey and guineafowl. I should think people had quite a job getting through the immensely hard guineafowl shells, but the colour of the yolks, if free range, is something to be seen.

One last point, don't sell your produce too cheap. Guineafowl are a gourmet food and not just any old chicken, so the sales talk and the price need to reflect this as they would if you were selling pheasant, wild duck or trout for instance.

MARKETING

It is no good producing a batch of guineafowl to sell and then ending up having to give them away or sell them at a knock down price.

Marketing means physically going to visit restaurants, hotels, pubs, local specialist butchers and farmers' markets. Avoid supermarkets as they pay a pittance, are slow to pay it, and what's more are quite likely to suddenly stop taking agreed deliveries.

Before you start marketing you should familiarise yourself thoroughly with your product. You must produce some pilot batches of birds, having killed, plucked and prepared them, which is a good way of learning about any problem areas in the production cycle. This way you will begin to get the feel of your product, and you'll have some samples to hand out into the bargain. Most chefs are familiar with guineafowl, but it will certainly help if you can provide a few recipes that you have tried out for yourself. There will be a lot to discuss such as how your birds are raised, what they are fed on, whether you have organic status and what kind of portions you will be selling. It's also useful if you can back it all up with some good photographs.

Guineafowl are a kind of halfway house between pheasants and chickens, but unlike pheasants they do not have a closed season. The other point is that as they are not shot there is no fear of finding shot in the meat. They taste very slightly gamey but nothing like as strong as pheasant or grouse, so even the most sensitive palate will not be put off.

Planning is vital. To begin with you should introduce your guineafowl onto the market with a special promotion and follow this up by supplying functions and parties. Remember, guineafowl are available all year round. If you keep them free range so that they are ready to be killed at 12 weeks, (or 7 weeks intensively) you would have a job to get four batches through a single house in a year, so you may need several houses in order to meet a regular demand. You will also have to consider the question of fresh or frozen birds.

The market for guineafowl in England is considerable according to the French who, in the year 2000, exported 619 tonnes of guineafowl meat into this country. You may find it better to link up with an organisation that sells to the London restaurants, where the consumer is perhaps more familiar with guineafowl than are people living outside the cities.

To sum up: the best sales, if you can get them, are into local restaurants, pubs and hotels, from the farm gate or at Farmers' Markets. You could also try local co-operatives which sell further afield, or specialist butchers and delicatessens. There will be a greater mark up on the former than the latter, but it may be best to have some of each.

HEALTH, DISEASES AND AILMENTS

When you read about the diseases and ailments that can affect guineafowl, you could be completely put off the idea of keeping them.

You must remember that these are outdoor birds, used to running round in flocks that can number several hundred. Put them indoors and, unless your management is first class, there are likely to be problems with houses containing 15,000 birds or more. Thus there are two types of disease problem, those found in intensively reared birds and those found in free range birds.

Let me say first of all that with a free range system your biggest problem is the fox. Guineafowl are extremely tough, they live a long time, 20 years or more, and rarely have any problems. The only thing I remember in many years of dealing with them is a female who had a bad case of frost bite in 1978. She lost all her toes, but lived on for several years, getting round on stumps instead of feet.

It is important to realise that throughout the rearing period, from day-old to the time when your birds are moved on either for killing or breeding, they are entirely dependant on us for food, water, light, ventilation and warmth. A breakdown in any of these systems will cause stress, one of the main killers of young birds, and one which is sometimes difficult to identify or measure. So sound management is absolutely essential if your birds are going to be free of disease.

Here are a few important points to remember when your birds arrive.

FOOD This must be the appropriate formulation for the age of your birds, it must be fresh and from a scheduled delivery, the size must be right, not too large and not too small and dusty. You must provide the right type of food container at the correct height to prevent wastage of food and disease. You must decide if you need an electric wire on the feeding line to stop the birds from perching on it.

WATER Your birds must have fresh water available at all times from the correct size drinkers. You must check regularly for leaks or faulty valves, and make sure that all doses of vitamins have been added at the appropriate times.

LIGHT You must organise the correct timing and wattage. Too little light and the birds are not stimulated to feed, too much or too bright and the birds become irritable and will start to scrap and feather peck.

HEATING AND VENTILATION These go hand in hand and the correct levels are crucial. Too much or too little of either and the birds will become stressed.

It is vital to maintain the correct temperature and flow of air because as the dung builds up the ammonia levels rise as well.

NOISE Always be careful not to make any sudden noises or movements in the vicinity of your birds which could cause them to panic and pile up in corners.

HYGIENE It is important to ensure that you are not bringing any diseases in from other sheds or farms or anywhere outside. To avoid this, you should arrange to have a Health Control Room that you go through when entering or leaving the building. As well as this, you must have an efficient vermin control programme in place to deal with rats and mice.

HEALTH PRECAUTIONS These are vitally important when you are rearing guineafowl intensively because disease can spread very quickly through large numbers of birds You need a room with an outside door and window, inside the entrance to the rearing shed that you can use as a Health Control Room, (see paragraph above on Hygiene). Here you can change boots, put on overalls and keep your records etc. You should divide the room with a low wooden barrier into roughly one third for dirty things, and two thirds for clean things. The dirty area should have a boot dip or disinfectant tray and pegs to hang outside clothing on; there should be washing facilities in the clean area. In warmer countries people put lime granules on the floor of the dirty area and on the paths outside leading to the shed.

Once you are inside and the door is closed you should take off your overalls and boots in the dirty area and step into the clean area. Here you put on the boots and overalls belonging to the building before proceeding into the rearing shed. This is really the minimum of precautions that you need to take. In the large breeding establishments where the rules are very strict, there are several seperate changing rooms and showers, different coloured wellingtons and complete changes of clothing.

The clean area of your Health Control Room must have a table and chair, and a basin with hot and cold water, soap and paper towels. Sometimes a fridge is required for vaccines, although it is more often than not used for storing the odd bottle or can of beer for hot days or visiting vets! You should also have a set of scales to check the birds' weights.

Anyone visiting the building who has had contact with other farms or poultry, or who is a vet, must cover him or herself completely with a disposable cap, mask and overalls, and plastic 'socks' to pull on over boots and trousers.

Needless to say, there are far fewer precautions necessary for free range birds. All you have to do before entering or leaving the area is wash your boots.

DISEASES We have divided up the diseases into four phases: 0 - 1 week, 1 - 4 weeks, 4 - 8 weeks intensive, and 4 - 12 weeks free range. This is only a brief look at diseases. It is not always easy to put your finger on what is wrong, so it is important to contact an avian vet if you are in doubt. Providing you have employed a proper sanitation, health and disinfectant programme between batches of birds, you should find that all will be well.

0 - 1 WEEK There should be very few problems during the first week which are not caused by bad management. The majority of deaths usually arise from 'starve-outs', 'dry-outs', die-outs', call them what you will, or smothering. 'Starve-outs' are chicks which never eat. They live off their yolk sac and once that is used up they die; the classic signs are withered legs. Smothering can occur if the chicks are too cold and crowd together to get warm. It can also happen if there is not enough light and they become disorientated and run about in a panic. Losses from smothering can be very high.

1 - 4 WEEKS At this stage the chicks should be feeding well and the wing feathers growing fast. Signs of good health are chicks sparing with each other, and plenty of skittish and energetic running about. You will sometimes see them 'crashed out' under the heat lamp looking rather dead but they soon get up when you touch them. Signs you certainly don't want to see are chicks standing about panting with drooping wings and heads, their wings looking longer than their bodies.

Chicks who are beginning to eat having finished their own reserves can go through a stage of feeling very uncomfortable; you may see them crouching on the ground, holding their wings away from their bodies and looking rather miserable. Most get through this period successfully but a few struggle and die. (In small-scale rearing finely chopped lettuce may come to the rescue!)

There can be a risk of **Salmonella**, (of which there are many strains, see later) or **Aspergillosis** which is normally caused by a leaky drinker. The floor litter becomes damp and produces fungal spores which affect the chicks' breathing. Stale food can also be the cause. The symptoms are birds stretching their necks to breathe and dying within 24 hours. To treat Aspergillosis you must clear up all the damp and infected areas and spray with a preventative disinfectant, before putting down fresh litter and replacing the food and water. Smothering can still be a risk at this stage.

Deformities. It is always worrying to see an otherwise perfectly healthy bird which has a stiff or splayed leg. The reasons for this are complex and I'm not sure that anyone knows exactly why it happens. It is called Perosis, and you will have to cull the bird.

Feather pecking. At this stage the body feathers start to grow, and if the birds are stressed for any reason such as being too hot or too cold, having too much light or incorrect or unbalanced food, the feathers become a target. The shoulders or wings will be attacked and then the backs and tails. This can lead to death.

If your birds are lethargic or off their food and dying for no apparent reason, you should gather up 6 to 12 of them and take them as soon as possible to your local avian laboratory for analysis. Birds that have just died or are dying are the best specimens to take.

4 - 8 WEEKS INTENSIVE The birds are far more prone to disease at this age. The shed is beginning to fill up with dust, the manure content in the floor litter is building up, and the density of birds is increasing, so all in all, a very susceptible environment. This is why it is so important to have good health precautions in place at all sheds and farms.

We are now looking at birds that are feathered up and growing very rapidly. They will be quite nervous and very skittish and noisy.

Apart from the odd deformity, some smothering and feather pecking, the main diseases will be bacterial and fungal.

The most common bacterial diseases are Salmonella and Mycoplasmosis. **Salmonella**. This can affect chicks and has many different strains and therefore needs to be analysed by an avian laboratory to determine which one is present. The main symptom is a whitish pasty discharge from the vent, and the bird will stand motionless and cheep incessantly. Salmonella is caused by poor hygiene and stale or infected food stuffs. It is brought in by rats and mice, and / or wild birds.

Mycoplasmosis / Infectious sinusitus. These two diseases are very similar, and although not often seen in the rearing shed, are quite unmistakeable when they do appear. The young birds sneeze and there is a discharge from the eyes and nostrils. In advanced cases, swellings appear above and below the eyes as well. If you are in any doubt about a diagnosis, the smell of the discharge is quite unmistakeable and quite awful! These diseases can be cleared up by one of the oxytetracyclin drugs.

Aspergillosis is the fungal disease that could affect your birds at this stage. It has already been described in the section on '1 - 4 weeks'.

4 - 12 WEEKS FREE RANGE This period of the bird's life brings it up to adulthood; you may even find a few eggs in the rearing shed at the end of this stage.

Free range birds usually fare much better than intensively reared ones. Although they need an extra month to reach the size and weight required by the abbatoir, they will do very well as long as the rearing paddock is clean and large enough. There is no doubt that they benefit greatly from the sunshine and the opportunity to find insects and have dustbaths. They will run around the paddock in great flocks, very noisy and nervous if there is an intruder.

There are big differences between free range conditions in England and in France, as the climate and rainfall vary a lot. 'Free range' in France means that birds have access to a dry pasture, whereas in this country it could mean the complete opposite, depending on where you are and what kind of soil you have. Moisture and damp will give rise to diseases such as Coccidiosis, Trichomoniasis and Hexamitiasis. There could be problems with worms as well.

Coccidiosis is normally seen in the spring or autumn months, but can also strike in the summer if the weather is wet. It is a protozoa of which there are 34 types. The birds have milky white diarrhoea, are very thirsty and usually die at night. There are several water-based drugs available to treat this, and a useful tip is to keep the grass short in the runs.

Trichomoniasis and **Hexamitiasis** are both sometimes associated with long spells of wet weather or damp, stale ground. They are caused by whip-like flagellae that live mainly in the caecal canal and small intestine. The eggs produced by these organisms are spread only too easily on boots and shoes. The symptoms of these two diseases are unthrifty birds with a foamy yellow diarrhoea. Mortality can be high. There are modern drugs available which can help, but it is still essential to put your birds onto clean ground and have a rigorous disinfection programme for all your drinkers, feeders and houses.

Worms including Gapes. More often than not worms come from using stale or over-used grassland or runs. The birds appear hungry and lazy but if picked up feel very light and thin. In the case of Gapes they seem to be trying to spit. The Gape worm anchores itself in the windpipe and gradually suffocates the bird. Luckily all worms can be cleared up with Flubenvet, (Flubendazole), quite easily. This is a white powder which is added to the birds' food.

Avian Flu. This is a comparatively new disease in Europe which has arrived from the Far East. It is said to be spread by wild birds, but the evidence for this is slightly lacking; it is more likely spread by poor bio-security. The guidelines and goal posts are being moved constantly, and therefore we are not able to offer any firm advice about this, except that the best bio-security is essential at all times. This is particularly important during visits from vets, representatives and Government officials who are travelling from farm to farm.

This by no means covers all the diseases that guineafowl can be prone to, but if your hygiene is good you shouldn't have any problems because, as I've mentioned before, guineafowl are quite resistant to most ailments. There could be a potential problem however, with the close breeding that is being carried out these days in quite unnatural and clinical conditions: there is a possibility that the natural hardiness of these birds is being eroded, and this is something that we must all be aware of.

This book would not have been possible without the kind assistance of the following organisations.

CIP Comité Interprofessionnel de la Pintade

Technopole Atalante-Champeaux, CS14226- 35042 Rennes, France

GALOR-France   Souvigny-de-Touraine, B.P. 142-37401 Amboise. France

SAVEL

BP 20 Sainte Sebastien
29870 Lannilis, France.

GUINEA FARM, 21357 White Pine Lane, New Vienna, IA 52065 USA.

ukguineafowl.com c|o S.S. BOUNDY, Partridge Meadows, Partridge Walls, Wembworthy, Chulmleigh. Devon. EX18 7SQ

44

INDEX